Ben Stacy Jerrik (Ed.)

Uinsky District

Ben Stacy Jerrik (Ed.)

Uinsky District

Perm Krai, Russia, Krai, Subdivisions of Russia, Administrative centre

Part Press

Imprint

Publisher:
Part Press is a trademark of
International Book Market Service Ltd., 17 Rue Meldrum, Beau Bassin, 1713-01 Mauritius
Email: info@bookmarketservice.com
Website: www.bookmarketservice.com

Published in 2013

Printed in: U.S.A., U.K., Germany. This book was not produced in Mauritius.

ISBN: 978-620-1-56468-8

Contents

Articles

References

Will_Todd

Will Todd (born 14 January 1970) is an English classical composer and pianist.

Biography

Todd was born in County Durham, attended Durham School and joined the choir of St Oswald's Church, Durham under its choirmaster David Higgins. He went on to study music at the University of Bristol. He is an accomplished jazz pianist and regularly performs with his trio, and this played a large role in one of his most famous works, his *Mass in Blue*.

Mass in Blue (originally entitled Jazz Mass) was commissioned by Hertfordshire Chorus [1] and first performed at the Cambridge Corn Exchange in July 2003 with Will Todd at the piano. Todd's wife, Bethany Halliday, performed the soprano solo alongside the Blue Planet Orchestra and Hertfordshire Chorus, conducted by David Temple.

His *The Blackened Man* won second prize in the *International Verdi Opera Competition* in 2002 and was later staged at the 2004 Buxton Festival. *The Screams of Kitty Genovese*, a music theatre work, has been produced at the Boston Conservatory and New York Musical Theatre Festival. His oratorio *Saint Cuthbert*, with a libretto by Ben Dunwell based on the life of the saint, has been performed and recorded by the Hallé Orchestra and Choir under Christopher Austin.

Among Angels was commissioned by the Genesis Foundation and was premiered by The Sixteen in Salzburg. Written for WNO Max, *Sweetness and Badness* was performed in a tour during autumn 2006, and *Whirlwind*, commissioned by Streetwise Opera, was first performed at The Sage Gateshead.

Works

- *Winter Dances* - orchestra (1989)
- *Midwinter* - Cantata for soprano and baritone, chorus and orchestra (1992)
- *Saint Cuthbert* - Oratorio for soprano, tenor and baritone, chorus, organ and orchestra (1995)
- *The Burning Road* - Cantata for soprano and baritone, chorus and orchestra (1996)
- *The Screams of Kitty Genovese* - 12 actor-singers and pit ensemble (1999)
- *A Song of Creation* - Cantata for choir, chamber choir, children's choir, soprano, tenor and bass soloists, organ and orchestra (2000)
- *The Blackened Man* - Opera with 7 roles, chorus and full orchestra or chamber ensemble (2001)
- *Mass in Blue* - soprano, chorus and jazz ensemble or trio (2003)
- *Gala and Gloria* - soprano, mezzo and baritone soloists, chorus, orchestra and brass band (2004)
- *Among Angels* - 16 part SATB with 2 harps (2006)
- *Sweetness and Badness* - Opera with 5 actor singers and small ensemble (2006)
- *Whirlwind* - Opera with 4 actor singers, mixed chorus, small ensemble and live sound score (2006)
- *Let us be True* - SATB, solo violin, piano, organ, percussion & strings. A setting of "Dover Beach" by Matthew Arnold (2006)
- *You have seen the House Built* - SATB choir and organ, commissioned by Chichester Cathedral for their 900th anniversary (2008)
- *Requiem* - soprano, electric guitar and chorus, commissioned by the Fairhaven Singers (2009)
- *Te Deum* - solo soprano, young people's choir (up to 3 parts), SATB chorus and instrumental ensemble. Latin text with additional words by Ben Dunwell (2009)
- *Clarinet Concerto* - Jazz clarinet and orchestra, premiered by Emma Johnson (2009)
- *The Call of Wisdom* [2] - SATB choir and organ, commissioned as part of the Diamond Jubilee of Elizabeth II to be sung by the Diamond Choir [3] (2012)

- *Songs of Love* [4] - SATB choir and Jazz trio, commissioned by Symphony Silicon Valley Chorale for their Silver Anniversary concert (2012)

External links

- Will Todd's homepage [5]
- press release: Oxford University Press: Diamond Jubilee anthem by Will Todd published [6] with links to purchase the score of *The Call of Wisdom* piece.

References

[1] http://www.hertfordshirechorus.org.uk/
[2] http://www.diamondchoir.co.uk/music.asp
[3] http://www.diamondchoir.co.uk/meet-the-choir.asp
[4] http://www.willtodd.com/workdetail.php?workID=87
[5] http://www.willtodd.com
[6] http://www.oup.com/uk/pressreleases/willtodd/

Classical_music

Classical music is the art music produced in, or rooted in, the traditions of Western music (both liturgical and secular), encompassing a broad period from roughly the 11th century to present times.[1] The central norms of this tradition became codified between 1550 and 1900, which is known as the common practice period.

European music is largely distinguished from many other non-European and popular musical forms by its system of staff notation, in use since about the 16th century.[2] Western staff notation is used by composers to prescribe to the performer the pitch, speed, meter, individual rhythms and exact execution of a piece of music. This leaves less room for practices such as improvisation and *ad libitum* ornamentation, which are frequently heard in non-European art music and popular music.[3] [4] [5]

Montage of some great classical music composers. From left to right:
Top row: Antonio Vivaldi, Johann Sebastian Bach, George Frideric Handel, Wolfgang Amadeus Mozart, Ludwig van Beethoven;
second row: Gioachino Rossini, Felix Mendelssohn, Frédéric Chopin, Richard Wagner, Giuseppe Verdi;
third row: Johann Strauss II, Johannes Brahms, Georges Bizet, Pyotr Ilyich Tchaikovsky, Antonín Dvořák;
bottom row – Edvard Grieg, Edward Elgar, Sergei Rachmaninoff, George Gershwin, Aram Khachaturian

The term "classical music" did not appear until the early 19th century, in an attempt to "canonize" the period from Johann Sebastian Bach to Beethoven as a golden age.[6] The earliest reference to "classical music" recorded by the *Oxford English Dictionary* is from about 1836.[1] []

Characteristics

Given the extremely broad variety of forms, styles, genres, and historical periods generally perceived as being described by the term "classical music," it is difficult to list characteristics that can be attributed to all works of that type. Vague descriptions are plentiful, such as describing classical music as anything that "lasts a long time," a statement made rather moot when one considers contemporary composers who are described as classical; or music that has certain instruments like violins, which are also found in other genres. However, there are characteristics that classical music contains that few or no other genres of music contain.[7]

Literature

The most outstanding characteristic of classical music is that the repertoire tends to be written down in musical notation, creating a musical part or score. This score typically determines details of rhythm, pitch, and, where two or more musicians (whether singers or instrumentalists) are involved, how the various parts are coordinated. The written quality of the music has, in addition to preserving the works, enabled a high level of complexity within them: Bach's fugues, for instance, achieve a remarkable marriage of boldly distinctive melodic lines weaving in counterpoint yet creating a coherent harmonic logic that would be impossible in the heat of live improvisation.[8]

Instrumentation

The instruments used in most classical music were largely invented before the mid-19th century (often much earlier), and codified in the 18th and 19th centuries. They consist of the instruments found in an orchestra, together with a few other solo instruments (such as the piano, harpsichord, and organ). The symphony orchestra is the most widely known medium for classical music.[] The orchestra includes members of the string, woodwind, brass, and percussion families.

The Dublin Philharmonic Orchestra performs Tchaikovsky's Fourth Symphony.

Electric instruments such as the electric guitar and the ondes Martenot appear occasionally in the classical music of the 20th and 21st centuries. Both classical and popular musicians have experimented in recent decades with electronic instruments such as the synthesizer, electric and digital techniques such as the use of sampled or computer-generated sounds, and the sounds of instruments from other cultures such as the gamelan.

None of the bass instruments existed until the Renaissance. In Medieval music, instruments are divided in two categories: loud instruments for use outdoors or in church, and quieter instruments for indoor use. The Baroque orchestra consisted of flutes, oboes, horns and violins, occasionally with trumpets and timpani.[] Many instruments today associated with popular music filled important roles in early classical music, such as bagpipes, vihuelas, hurdy-gurdies, and some woodwind instruments. On the other hand, instruments such as the acoustic guitar, once associated mainly with popular music, gained prominence in classical music in the 19th and 20th centuries.

While equal temperament became gradually accepted as the dominant musical temperament during the 18th century, different historical temperaments are often used for music from earlier periods. For instance, music of the English Renaissance is often performed in meantone temperament. Keyboards almost all share a common layout (often called the piano keyboard).

Form

Whereas most popular styles lend themselves to the song form, classical music has been noted for its development of highly sophisticated forms of instrumental music:[9] these include the concerto, symphony, sonata, suite, étude, symphonic poem, opera, and others.

Classical composers often aspire to imbue their music with a very complex relationship between its affective (emotional) content and the intellectual means by which it is achieved. Many of the most esteemed works of classical music make use of musical development, the process by which a musical idea or motif is repeated in different contexts or in altered form. The sonata form and fugue employ rigorous forms of musical development.

Technical execution

Along with a desire for composers to attain high technical achievement in writing their music, performers of classical music are faced with similar goals of technical mastery, as demonstrated by the proportionately high amount of schooling and private study most successful classical musicians have had when compared to "popular" genre musicians, and the large number of secondary schools, including conservatories, dedicated to the study of classical music.

Complexity

Professional performance of classical music repertoire demands a significant level of proficiency in sight-reading and ensemble playing, thorough understanding of tonal and harmonic principles, knowledge of performance practice, and a familiarity with the style/musical idiom inherent to a given period, composer or musical work are among the most essential of skills for the classically trained musician.

Works of classical repertoire often exhibit artistic complexity through the use of counterpoint, thematic development, phrasing, harmonization, modulation (change of key), texture, and, of course, musical form itself. Larger-scale compositional forms (such as that of the symphony, concerto, opera or oratorio, for example) usually represent a hierarchy of smaller units consisting of phrases, periods, sections, and movements. Musical analysis of a composition aims at achieving greater understanding of it, leading to more meaningful hearing and a greater appreciation of the composer's style.

Society

Classical music regularly features as background music for movies, television programmes, advertisements and events. "Nessun dorma" from Giacomo Puccini's opera *Turandot* for example was the theme tune for the 1990 FIFA World Cup.

History

The major time divisions of classical music are the early music period, which includes Medieval (500–1400) and Renaissance (1400–1600), the Common practice period, which includes the Baroque (1600–1750), Classical (1750–1830) and Romantic (1804–1949) periods, and the modern and contemporary period, which includes 20th century (1900–2000) and contemporary (1975–current).

The dates are generalizations, since the periods overlapped and the categories are somewhat arbitrary. For example, the use of counterpoint and fugue, which is considered characteristic of the Baroque era, was continued by Haydn, who is classified as typical of the Classical period. Beethoven, who is often described as a founder of the Romantic period, and Brahms, who is classified as Romantic, also used counterpoint and fugue, but other characteristics of their music define their period.

The prefix *neo* is used to describe a 20th-century or contemporary composition written in the style of an earlier period, such as Classical or Romantic. Stravinsky's *Pulcinella*, for example, is a neoclassical composition because it is stylistically similar to works of the Classical period.

Roots

The roots of Western classical music lie in early Christian liturgical music, and its influences date back to the Ancient Greeks. Development of individual tones and scales was done by ancient Greeks such as Aristoxenus and Pythagoras.[10] Pythagoras created a tuning system and helped to codify musical notation. Ancient Greek instruments such as the aulos (a reed instrument) and the lyre (a stringed instrument similar to a small harp) eventually led to the modern-day instruments of a classical orchestra.[11] The antecedent to the early period was the era of ancient music from before the fall of the Roman Empire (476 AD). Very little music survives from this time, most of it from Ancient Greece.

Early period

The Medieval period includes music from after the fall of Rome to about 1400. Monophonic chant, also called plainsong or Gregorian chant, was the dominant form until about 1100.[12] Polyphonic (multi-voiced) music developed from monophonic chant throughout the late Middle Ages and into the Renaissance, including the more complex voicings of motets. The Renaissance period was from 1400 to 1600. It was characterized by greater use of instrumentation, multiple interweaving melodic lines, and the use of the first bass instruments. Social dancing became more widespread, so musical forms appropriate to accompanying dance began to standardize.

A musician plays the vielle in a fourteenth-century Medieval manuscript.

It is in this time that the notation of music on a staff and other elements of musical notation began to take shape.[13] This invention made possible the separation of the composition of a piece of music from its *transmission*; without written music, transmission was oral, and subject to change every time it was transmitted. With a musical score, a work of music could be performed without the composer's presence.[12] The invention of the movable-type printing press in the 15th century had far-reaching consequences on the preservation and transmission of music.[14]

Typical stringed instruments of the Early Period include the harp, lute, vielle, and psaltery, while wind instruments included the flute family (including recorder), shawm (an early member of the oboe family), trumpet, and the bagpipes. Simple pipe organs existed, but were largely confined to churches, although there were portable varieties.[15] Later in the period, early versions of keyboard instruments like the clavichord and harpsichord began to appear. Stringed instruments such as the viol had emerged by the 16th century, as had a wider variety of brass and reed instruments. Printing enabled the standardization of descriptions and specifications of instruments, as well as instruction in their use.[16]

The chanson *Belle, bonne, sage* by Baude Cordier, an *ars subtilior* piece included in the Chantilly Codex.

Johannes Ockeghem, Kyrie "Au travail suis," excerpt

Common practice period

The common practice period is when many of the ideas that make up western classical music took shape, standardized, or were codified. It began with the Baroque era, running from roughly 1600 to the middle of the 18th century. The Classical era followed, ending roughly around 1820. The Romantic era ran through the 19th century, ending about 1910.

Baroque music

Baroque music is characterized by the use of complex tonal counterpoint and the use of a basso continuo, a continuous bass line. Music became more complex in comparison with the songs of earlier periods.[] The beginnings of the sonata form took shape in the canzona, as did a more formalized notion of theme and variations. The tonalities of major and minor as means for managing dissonance and chromaticism in music took full shape.[17]

Baroque instruments including hurdy gurdy, harpsichord, bass viol, lute, violin, and baroque guitar

During the Baroque era, keyboard music played on the harpsichord and pipe organ became increasingly popular, and the violin family of stringed instruments took the form generally seen today. Opera as a staged musical drama began to differentiate itself from earlier musical and dramatic forms, and vocal forms like the cantata and oratorio became more common.[18] Vocalists began adding embellishments to melodies.[] Instrumental ensembles began to distinguish and standardize by size, giving rise to the early orchestra for larger ensembles, with chamber music being written for smaller groups of instruments where parts are played by individual (instead of massed) instruments. The concerto as a vehicle for solo performance accompanied by an orchestra became widespread, although the relationship between soloist and orchestra was relatively simple. The theories surrounding equal temperament began to be put in wider practice, especially as it enabled a wider range of chromatic possibilities in hard-to-tune keyboard instruments. Although Bach did not use equal temperament, as a modern piano is generally tuned, changes in the temperaments from the meantone system, common at the time, to various temperaments that made modulation between all keys musically acceptable, made possible Bach's Well-Tempered Clavier.[19]

Classical period music

The Classical period, from about 1750 to 1820, established many of the norms of composition, presentation, and style, and was also when the piano became the predominant keyboard instrument. The *basic* forces required for an orchestra became somewhat standardized (although they would grow as the potential of a wider array of instruments was developed in the following centuries). Chamber music grew to include ensembles with as many as 8 to 10 performers for serenades. Opera continued to develop, with regional styles in Italy, France, and German-speaking lands. The *opera buffa*, a form of comic opera, rose in popularity. The symphony came into its own as a musical form, and the concerto was developed as a vehicle for displays of virtuoso playing skill. Orchestras no longer required a harpsichord (which had been part of the traditional *continuo* in the Baroque style), and were often led by the lead violinist (now called the concertmaster).[20]

Joseph Haydn (1732–1809) c. 1770

Wind instruments became more refined in the Classical period. While double reeded instruments like the oboe and bassoon became somewhat standardized in the Baroque, the clarinet family of single reeds was not widely used until Mozart expanded its role in orchestral, chamber, and concerto settings.

Romantic era music

The music of the Romantic era, from roughly the first decade of the 19th century to the middle 20th century, was characterized by increased attention to an extended melodic line, as well as expressive and emotional elements, paralleling romanticism in other art forms. Musical forms began to break from the Classical era forms (even as those were being codified), with free-form pieces like nocturnes, fantasias, and preludes being written where accepted ideas about the exposition and development of themes were ignored or minimized.[21] The music became more chromatic, dissonant, and tonally colorful, with tensions (with respect to accepted norms of the older forms) about key signatures increasing.[22] The art song (or *Lied*) came to maturity in this era, as did the epic scales of grand opera, ultimately transcended by Richard Wagner's Ring cycle.[23]

In the 19th century, musical institutions emerged from the control of wealthy patrons, as composers and musicians could construct lives independent of the nobility. Increasing interest in music by the growing middle classes throughout western Europe spurred the creation of organizations for the teaching, performance, and preservation of music. The piano, which achieved its modern construction in this era (in part due to industrial advances in metallurgy) became widely popular with the middle class, whose demands for the instrument spurred a large number of piano builders. Many symphony orchestras date their founding to this era.[22] Some musicians and composers were the stars of the day; some, like Franz Liszt and Niccolò Paganini, fulfilled both roles.[24]

The family of instruments used, especially in orchestras, grew. A wider array of percussion instruments began to appear. Brass instruments took on larger roles, as the introduction of rotary valves made it possible for them to play a wider range of notes. The size of the orchestra (typically around 40 in the Classical era) grew to be over 100.[22] Gustav Mahler's 1906 *Symphony No. 8*, for example, has been performed with over 150 instrumentalists and choirs of over 400.

European cultural ideas and institutions began to follow colonial expansion into other parts of the world. There was also a rise, especially toward the end of the era, of nationalism in music (echoing, in some cases, political sentiments of the time), as composers such as Edvard Grieg, Nikolai Rimsky-Korsakov, and Antonín Dvořák echoed traditional music of their homelands in their compositions.[25]

20th-century, modern, and contemporary music

Modernism (1905–1985) marked a period when many composers rejected certain values of the common practice period, such as traditional tonality, melody, instrumentation, and structure. Composers, academics, and musicians developed extensions of music theory and technique. 20th century classical music, encompassing a wide variety of post-Romantic styles composed through the year 1999, includes late Romantic, Modern and Postmodern styles of composition. The term "contemporary music" is sometimes used to describe music composed in the late 20th century through to the present day.

Igor Stravinsky and Pablo Picasso collaborated on *Pulcinella* in 1920.

Significance of written notation

Modernist view of the significance of the score

The modernist views hold that classical music is considered primarily a *written* musical tradition, preserved in music notation, as opposed to being transmitted orally, by rote, or by recordings of particular performances. While there are differences between particular performances of a classical work, a piece of classical music is generally held to transcend any interpretation of it. The use of musical notation is an effective method for transmitting classical music, since the written music contains the technical instructions for performing the work.

The written score, however, does not usually contain explicit instructions as to how to interpret the piece in terms of production or performance, apart from directions for dynamics, tempo and expression (to a certain extent). This is left to the discretion of the performers, who are guided by their personal experience and musical education, their knowledge of the work's idiom, their personal artistic tastes, and the accumulated body of historic performance practices.

Criticism of the modernist view

Some critics express the opinion that it is only from the mid 19th century, and especially in the 20th century, that the score began to hold such a high significance. Previously, improvisation, rhythmic flexibility, improvisatory deviation from the score and oral tradition of playing was integral to style of music. Yet in the 20th century, this oral tradition and passing on of stylistic features within classical music disappeared. Instead, musicians use the score to play music, yet even given the score, there is considerable controversy about how to perform the works. Some of this controversy relates to the fact that this score-centric approach has led to performing styles that emphasize metrically strict block-rhythms (just as the music is notated in the score).

Some quotes that highlight this criticism of modernist overvaluing of the score:

- "... one of the most stubborn modern misconceptions concerning baroque music is that a metronomic regularity was intended" (Baroque Interpretation in *Grove* 5th edition by Robert Donington)
- "The history of this particular idea is littered with dead ends and failed projects. It is high time these misconceptions are addressed with academic rigour." *History of Metaphysics* by Andrew Pyle
- "Too many teachers, conditioned to 20th Century ideas, teach Bach and other Baroque music exactly the wrong way. This leads to what musicologist Sol Babitz calls 'sewing machine Bach'."[26]
- "... tendency to look alike, sound alike and think alike. The conservatories are at fault and they have been at fault for many years now. Any sensitive musician going around the World has noted the same thing. The conservatories, from Moscow and Leningrad to Juilliard, Curtis and Indiana, are producing a standardized product.

[...] clarity, undeviating rhythm, easy technique, 'musicianship'. I put the word musicianship in quotes, because as often as not, it is a false kind of musicianship – a musicianship that sees the tree and not the forest, that takes care of the detail but ignores the big picture; a musicianship that is tied to the printed note rather than to emotional meaning of a piece.
The fact remains that there is a dreadful uniformity today and also an appalling lack of knowledge about the culture and performance traditions of the past." ("Music Schools Turning out Robots?"[26] by Harold C. Schonberg)

Improvisation

Improvisation once played an important role in classical music. A remnant of this improvisatory tradition in classical music can be heard in the cadenza, a passage found mostly in concertos and solo works, designed to allow skilled performers to exhibit their virtuoso skills on the instrument. Traditionally this was improvised by the performer; however, it is often written for (or occasionally by) the performer beforehand. Improvisation is also an important aspect in authentic performances of operas of Baroque era and of bel canto (especially operas of Vincenzo Bellini), and is best exemplified by the *da capo aria*, a form by which famous singers typically perform variations of the thematic matter of the aria in the recapitulation section ('B section' / the 'da capo' part). An example is Beverly Sills' complex, albeit pre-written, variation of *Da tempeste il legno infranto* from Händel's *Giulio Cesare*.

Its written transmission, along with the veneration bestowed on certain classical works, has led to the expectation that performers will play a work in a way that realizes in detail the original intentions of the composer. During the 19th century the details that composers put in their scores generally increased. Yet the opposite trend – admiration of performers for new "interpretations" of the composer's work – can be seen, and it is not unknown for a composer to praise a performer for achieving a better realization of the original intent than the composer was able to imagine. Thus, classical performers often achieve very high reputations for their musicianship, even if they do not compose themselves. Generally however, it is the composers who are remembered more than the performers.

The primacy of the composer's written score has also led, today, to a relatively minor role played by improvisation in classical music, in sharp contrast to the practice of musicians who lived during the baroque, classical and romantic era. Improvisation in classical music performance was common during both the Baroque era and in the nineteenth, yet lessened strongly during the second half of the 19th and in the 20th centuries. Recently the performance of such music by modern classical musicians has been enriched by a revival of the old improvisational practices. During the classical period, Mozart and Beethoven often improvised the cadenzas to their piano concertos (and thereby encouraged others to do so), but they also provided written cadenzas for use by other soloists. In opera, the practice of singing strictly by the score i.e. *come scritto*, is famously propagated by Maria Callas, who called this practice 'straitjacketing' and implied that it allows the intention of the composer to be understood better, especially during studying the music for the first time.

Relationship to other music traditions

Popular music

Classical music has often incorporated elements or material from popular music of the composer's time. Examples include occasional music such as Brahms' use of student drinking songs in his *Academic Festival Overture*, genres exemplified by Kurt Weill's *The Threepenny Opera*, and the influence of jazz on early- and mid-20th-century composers including Maurice Ravel, exemplified by the movement entitled "Blues" in his sonata for violin and piano.[27] Certain postmodern, minimalist and postminimalist classical composers acknowledge a debt to popular music.[28]

Numerous examples show influence in the opposite direction, including popular songs based on classical music, the use to which *Pachelbel's Canon* has been put since the 1970s, and the musical crossover phenomenon, where classical musicians have achieved success in the popular music arena.[29]

Folk music

Composers of classical music have often made use of folk music (music created by musicians who are commonly not classically trained, often from a purely oral tradition). Some composers, like Dvořák and Smetana,[] have used folk themes to impart a nationalist flavor to their work, while others (like Bartók) have used specific themes lifted whole from their folk-music origins.[]

Commercialism

Certain staples of classical music are often used commercially (either in advertising or in movie soundtracks). In television commercials, several passages have become clichéd, particularly the opening of Richard Strauss' *Also sprach Zarathustra* (made famous in the film *2001: A Space Odyssey*) and the opening section "O Fortuna" of Carl Orff's *Carmina Burana*, often used in the Horror genre to denote demonic powers; other examples include the *Dies Irae* from the Verdi Requiem, Edvard Grieg's *In the Hall of the Mountain King* from *Peer Gynt*, the opening bars of Beethoven's Symphony No. 5, Wagner's "Ride of the Valkyries" from *Die Walküre*, Rimsky-Korsakov's "Flight of the Bumblebee", and excerpts of Aaron Copland's *Rodeo*. Shawn Vancour argues that the commercialization of classical music in the early 20th century served to harm the music industry through inadequate representation.[30]

Similarly, movies and television often revert to standard, clichéd snatches of classical music to convey refinement or opulence: some of the most-often heard pieces in this category include Mozart's *Eine kleine Nachtmusik*, Vivaldi's *Four Seasons*, Mussorgsky's *Night on Bald Mountain* (as orchestrated by Rimsky-Korsakov), and Rossini's *William Tell Overture*.

Education

Throughout history, parents have often made sure that their children receive classical music training from a young age. Some parents pursue music lessons for their children for social reasons or in an effort to instill a sense of self-discipline. Some believe that knowledge of important works of classical music is part of a good general education.

During the 1990s, several research papers and popular books wrote on what came to be called the "Mozart effect": an observed temporary, small elevation of scores on certain tests as a result of listening to Mozart's works. The approach has been popularized in a book by Don Campbell, and is based on an experiment published in *Nature* suggesting that listening to Mozart temporarily boosted students' IQ by 8 to 9 points.[31] This popularized version of the theory was expressed succinctly by the *New York Times* music columnist Alex Ross: "researchers... have determined that listening to Mozart actually makes you smarter."[32] Promoters marketed CDs claimed to induce the effect. Florida passed a law requiring toddlers in state-run schools to listen to classical music every day, and in 1998 the governor of Georgia budgeted $105,000 per year to provide every child born in Georgia with a tape or CD of classical music. One of the co-authors of the original studies of the Mozart effect commented "I don't think it can hurt. I'm all for exposing children to wonderful cultural experiences. But I do think the money could be better spent on music education programs."[33]

In 1996–1997, a research study was conducted on a large population of middle age students in the Cherry Creek School District in Denver, Colorado, USA. The study showed that students who actively listen to classical music before studying had higher academic scores. The research further indicated that students who listened to the music prior to an examination also had positively elevated achievement scores. Students who listened to rock-and-roll or country had moderately lower scores. The study further indicated that students who used classical during the course of study had a significant leap in their academic performance; whereas, those who listened to other types of music had significantly lowered academic scores. The research was conducted over several schools within the Cherry

Creek School District and was conducted through University of Colorado. This study is reflective of several recent studies (i.e. Mike Manthei and Steve N. Kelly of the University of Nebraska at Omaha; Donald A. Hodges and Debra S. O'Connell of the University of North Carolina at Greensboro; etc.) and others who had significant results through the discourse of their work.[34]

See also

- List of classical music composers by era
- List of classical and art music traditions

Nation specific:

- American classical music
- Australian classical music
- Canadian classical music
- French classical music
- Indian classical music
- Iranian classical music
- Italian classical music
- Ottoman classical music
- Russian classical music
- Classical music of the United Kingdom

Notes

[1] "Classical", *The Oxford Concise Dictionary of Music*, ed. Michael Kennedy, (Oxford, 2007), *Oxford Reference Online*. Retrieved July 23, 2007.

[2] Chew, Geffrey & Rastall, Richard. "Notation, §III, 1(vi): Plainchant: Pitch-specific notations, 13th–16th centuries", *Grove Music Online*, ed. L. Macy (accessed July 23 2007), grovemusic.com (http://www.grovemusic.com/) (subscription access).

[3] Malm, W.P./Hughes, David W.. "Japan, §III, 1: Notation systems: Introduction", *Grove Music Online*, ed. L. Macy (accessed July 23 2007), grovemusic.com (http://www.grovemusic.com/) (subscription access).

[4] IAN D. BENT, DAVID W. HUGHES, ROBERT C. PROVINE, RICHARD RASTALL, ANNE KILMER. "Notation, §I: General", *Grove Music Online*, ed. L. Macy (accessed July 23 2007), grovemusic.com (http://www.grovemusic.com/) (subscription access).

[5] Middleton, Richard. "Popular music, §I, 4: Europe & North America: Genre, form, style", *Grove Music Online*, ed. L. Macy (accessed July 23 2007), grovemusic.com (http://www.grovemusic.com/) (subscription access).

[6] Rushton, Julian, *Classical Music*, (London, 1994), 10

[7] Kennedy, Michael (2006), *The Oxford Dictionary of Music*, p. 178

[8] Knud Jeppesen: "Bach's music grows out of an ideally harmonic background, against which the voices develop with a bold independence that is often breath-taking." Quoted from Adele Katz (1946; reprinted 2007)

[9] Julian Johnson (2002) *Who Needs Classical Music?: Cultural Choice and Musical Value*: p. 63.

[10] Grout, p. 28

[11] Grout (1988)

[12] Grout, pp. 75–76

[13] Grout, p. 61

[14] Grout, pp. 175–176

[15] Grout, pp. 72–74

[16] Grout, pp. 222–225

[17] Grout, pp. 300–332

[18] Grout, pp. 341–355

[19] Grout, p. 378

[20] Grout, p. 463

[21] Swafford, p. 200

[22] Swafford, p. 201

[23] Grout, pp. 595–612

[24] Grout, p. 543

[25] Grout, pp. 634,641–2

[26] "Music Schools Turning out Robots?" (http://news.google.com/newspapers?nid=1873&dat=19691019&id=dpMeAAAAIBAJ&sjid=ncsEAAAAIBAJ&pg=874,4160516) by Harold C. Schonberg; *Daytona Beach Morning Journal* – October 19, 1969

[27] Kelly, Barbara. L. "Ravel, Maurice, §3: 1918–37", *Grove Music Online*, ed. L. Macy (accessed July 23 2007), grovemusic.com (http://www.grovemusic.com/) (subscription access).

[28] See, for example, Siôn, Pwyll Ap. "Nyman, Michael", *Grove Music Online*, ed. L. Macy (accessed July 23 2007), grovemusic.com (http://www.grovemusic.com/) (subscription access).

[29] Notable examples are the *Hooked on Classics* series of recordings made by the Royal Philharmonic Orchestra in the early 1980s and the classical crossover violinists Vanessa Mae and Catya Maré.

[31] Prelude or requiem for the 'Mozart effect'? (http://www.nature.com/nature/journal/v400/n6747/abs/400827a0.html) *Nature* **400** (August 26, 1999): 827.

[32] Ross, Alex. "Classical View; Listening To Prozac... Er, Mozart" (http://query.nytimes.com/gst/fullpage.html?res=9C05EFDF1239F93BA1575BC0A962958260), *The New York Times*, August 28, 1994. Retrieved on May 16, 2008.

[33] Goode, Erica. "Mozart for Baby? Some Say, Maybe Not" (http://query.nytimes.com/gst/fullpage.html?res=9C03E0D81131F930A3575BC0A96F958260&sec=&spon=&pagewanted=3), *The New York Times*, August 3, 1999. Retrieved on May 16, 2008.

References

- Johnson, Julian (2002), *Who Needs Classical Music?: Cultural Choice and Musical Value*. Oxford University Press, 140pp.
- Katz, Adele (1946; reprinted 2007), *Challenge to Musical Tradition – A New Concept of Tonality*. Alfred A. Knopf/reprinted by Katz Press, 444pp., ISBN 1-4067-5761-6.
- Kennedy, Michael (2006), *The Oxford Dictionary of Music*, 985 pages, ISBN 0-19-861459-4

Further reading

- Copland, Aaron (1957) *What to Listen for in Music*; rev. ed. McGraw-Hill. (paperback).
- --"-- (1988) --"--; with an introduction by William Schuman. McGraw-Hill ISBN 0-07-013091-4
- --"-- (2002) --"--; with a foreword and epilogue by Alan Rich; with an introduction by William Schuman. New American Library ISBN 0-451-52867-0 (reissued 2009 with new appreciation by Leonard Slatkin)
- Grout, Donald Jay; Palisca, Claude V. (1996) *A History of Western Music*, Fifth edition. W. W. Norton & Company. ISBN 0-393-96904-5 (hardcover).
- Hanning, Barbara Russano; Grout, Donald Jay (1998 rev. 2009) *Concise History of Western Music*. W. W. Norton & Company. ISBN 0-393-92803-9 (hardcover).
- Johnson, Julian (2002) *Who Needs Classical Music?: cultural choice and musical value*. Oxford University Press. ISBN 0-19-514681-6.
- Kamien, Roger (2008) *Music: an appreciation*; 6th brief ed. McGraw-Hill ISBN 978-0-07-340134-8
- Lihoreau, Tim; Fry, Stephen (2004) *Stephen Fry's Incomplete and Utter History of Classical Music*. Boxtree. ISBN 978-0-7522-2534-0
- Scholes, Percy Alfred; Arnold, Denis (1988) *The New Oxford Companion to Music*. Oxford University Press. ISBN 0-19-311316-3 (paperback).
- Sorce Keller, Marcello (2011) *What Makes Music European. Looking Beyond Sound*. Latham, NJ: Scarecrow Press (USA).
- Taruskin, Richard (2005, rev. Paperback version 2009) *Oxford History of Western Music*. Oxford University Press (USA). ISBN 978-0-19-516979-9 (Hardback), ISBN 978-0-19-538630-1 (Paperback)

External links

- Historical classical recordings from the British Library Sound Archive (http://sounds.bl.uk/BrowseCategory.aspx?category=Classical-music) (available only to users in the member countries of the European Union)
- Chronological list of recorded classical composers (http://www.discogs.com/lists/Classical-Composers/396)

Composer

A **composer** (Latin *com+ponere*, literally "one who puts together") is a person who creates music, either by musical notation or oral tradition, for interpretation and performance, or through direct manipulation of sonic material as electroacoustic music.

French composer Louis-Nicolas Clérambault composing at the piano

History

The level of distinction between composers and other musicians varies, which affects issues such as copyright and the deference given to individual interpretations of a particular piece of music. In the development of European classical music, the function of composing music initially did not have much greater importance than that of performing it. The preservation of individual compositions did not receive enormous attention and musicians generally had no qualms about modifying compositions for performance. Over time, however, the written notation of the composer came to be treated as strict instructions from which performers should not deviate without good practical or artistic reason. Performers do, however, play the music and interpret it in a way that is all their own. In fact, in the concerto form, the soloist would often compose and perform a cadenza as a way to express their individual interpretation of the piece.

In as much as the role of the composer in western art music has seen continued solidification, in alternative idioms (i.e. jazz, experimental music) it has in some ways become increasingly complex or vague. For instance, in certain contexts the line between composer and performer, sound designer, arranger, producer, and other roles, can be quite blurred.

The term "composer" is often used to refer to composers of instrumental music, such as those found in classical, jazz or other forms of art and traditional music. In popular and folk music, the composer is usually called a songwriter, since the music generally takes the form of a song.

Since the mid-20th century, the term has expanded to accommodate creators of electroacoustic music, in which composers directly create sonic material in any of the various electronic media. This is distinct from instrumental composition, where the work is represented by a musical score to be interpreted by performers.

Clustering

Famous composers have tended to cluster in certain cities throughout history. Based on over 12,000 prominent composers listed in Grove Music Online and using word count measurement techniques the most important cities for classical music can be quantitatively identified.[1]

Paris has been the main hub for classical music of all times. It was ranked fifth in the 15th and 16th centuries but first in the 17th to 20th centuries inclusive. London was the second most meaningful city: eight in the 15th century, seventh in the 16th, fifth in the 17th, second in the 18th and 19th centuries, and fourth in the 20th century. Rome topped the rankings in the 15th century, dropped to second in the 16th and 17th centuries, eight in the 18th century, ninth in the 19th century but back at sixth in the 20th century. Berlin appears in the top ten ranking only in the 18th century, and was ranked third most important city in both the 19th and 20th centuries. New York entered the rankings in the 19th century (at fifth place) and stood at second rank in the 20th century.

The patterns are very similar for a sample of 522 top composers.[2]

See also

- Lists of composers

References

- Merriam-Webster dictionary on 'composer', accessed 17 November 2010 (http://www.merriam-webster.com/dictionary/composer)

External links

- Composer societies (http://www.dmoz.org//Arts/Music/Organizations/By_Composer//) at the Open Directory Project
- Composers (http://www.dmoz.org//Arts/Music/Composition/Composers//) at the Open Directory Project
- Composers and arrangers (http://www.dmoz.org//Business/Arts_and_Entertainment/Music/Composers_and_Arrangers//) at the Open Directory Project

Pianist

A **pianist** (/ˈpiːənɪst/ ***PEE**-ə-nist*) is a musician who plays the piano. Most forms of Western music can make use of the piano. Consequently, pianists have a wide variety of repertoire and styles to choose from, including classical music, jazz, and all sorts of popular music. Most pianists can also play other keyboard instruments such as the synthesizer, harpsichord and the organ.

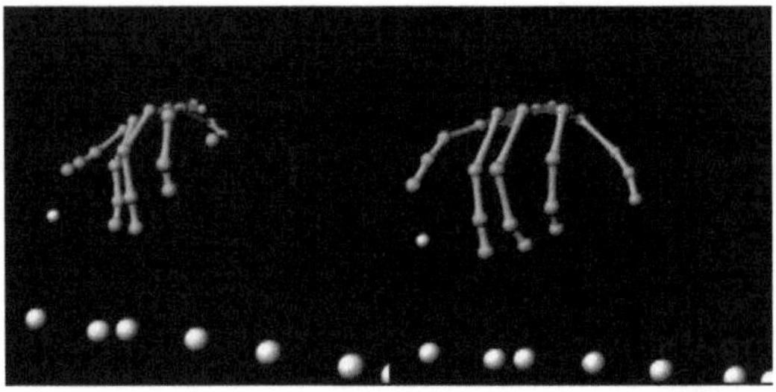

Motion capture of two pianists' fingers playing the same piece (slow motion, no sound).[1]

Choice of genres

Modern classical pianists dedicate their careers to learning new works/expanding their repertoire, teaching, research, performing and recording. They generally do not write or transcribe music as pianists did in the 19th century. Some might specialize in accompanying and playing chamber music while others (relatively few) will perform as full-time soloists.

Mozart could be considered the first 'concert pianist' as he performed widely on the piano. Beethoven and Liszt were also famed for their playing, however, as we do not have audio recordings of these pianists, we rely completely on written commentary to give us an account of their technique and style.

Jazz pianists almost always perform with other musicians. Their playing is freer than that of classical pianists and they create an air of spontaneity in their performances. They generally do not write down their compositions; improvisation is a significant part of their work. Well known Jazz pianists include Art Tatum, Duke Ellington, Oscar Peterson and Herbie Hancock.

Popular pianists might work as a session musician, teacher, arranger as well as a live performer. Notable pianists include Victor Borge who performed as a comedian, Richard Clayderman and Liberace, also a singer who at the height of his fame, was one of the highest paid entertainers in the world.

A single listing of pianists in all genres would be impractical, given the multitude of musicians noted for their performances on the instrument. Below are links to lists of well-known or influential pianists divided by genres:

Classical pianists

- List of classical pianists (recorded)
- List of classical pianists
- List of classical piano duos (performers)

Jazz pianists

- List of jazz pianists

Pop and rock music pianists

- List of pop and rock pianists

Blues pianists

- List of blues musicians
- List of boogie woogie musicians

Gospel pianists

- List of gospel musicians

New Age pianists

- List of New Age music artists

Pianist-composers

Many important composers were also virtuoso pianists. The following is an incomplete list of such musicians.

Classical period

- Ludwig van Beethoven
- Wolfgang Amadeus Mozart
- Johann Nepomuk Hummel
- Muzio Clementi
- Carl Maria von Weber

Romantic period

- Franz Liszt
- Frédéric Chopin
- Felix Mendelssohn
- Johannes Brahms
- Sergei Rachmaninoff
- Anton Rubinstein
- Camille Saint-Saëns
- Isaac Albéniz
- Nikolai Medtner
- Charles-Valentin Alkan
- Anton Arensky

Modern period

- Sergei Prokofiev
- Béla Bartók
- Claude Debussy
- Alexander Scriabin
- Dmitri Shostakovich
- Leonard Bernstein
- Ferruccio Busoni

References

[1] Error: Bad DOI specified: 10.1371/journal.pone.0050901

Durham_School

Durham School is an English independent school for pupils aged between 3 to 18 years. Founded by the Bishop of Durham, Thomas Langley in 1414 as the Durham Grammar School, it received royal foundation by King Henry VIII in 1541 following the Dissolution of the Monasteries during the Protestant Reformation. It is the city's oldest institution of learning.

The School is located in Durham, North East England and was an all-boys institution until becoming fully coeducational in 1985. A member of The Headmasters' and Headmistresses' Conference, it enrolls 650 day and boarding students. Its preparatory institution, known as Bow, Durham School, enrolls a further 160 pupils. Durham and Bow have a long list of distinguished former pupils including politicians, clergy and British aristocracy. Former students are known as *Old Dunelmians*.[1]

History

The history of Durham School can be divided into three sections. Firstly there is the time from its founding by Langley in 1414, then in 1541 Henry VIII refounded it and finally in 1844 the school moved from its site on Palace Green to its current location. The school is often referred to in histories and the Oxford Dictionary of National Biography as "Durham Grammar School".

To 1541

Durham School was founded by Thomas Langley[] in 1414, which was the foundation date accepted by the Clarendon Commission into public schools in 1861, making it the 18th oldest in Britain. It is possible, however, that it may actually have origins in the Priory at Lindisfarne, being moved to Durham City to escape marauding Viking invaders around the time that St. Cuthbert's body was brought to what is now the site of Durham Cathedral. The school was in Langley's time situated on the east side of Palace Green to the north of the cathedral.

1541 to 1844

At the time of the Dissolution of the Monasteries during the Protestant Reformation in 1541, the school was refounded by Henry VIII. It remained in the same location, indeed the Headmaster Henry Stafford remained in post, during this period. The School had some interesting times - for example, in 1640 the "old School Buildings were utterly destroyed by the Scots, and the Head Master [Richard Smelt] retired to his Rectory of Easingwold where he took pupils"[2] and the School was "recognised and endowed by the Parlamentary Commission 1652-3".[3] Homeless due to the burning down of its buildings, the school continued in various houses in the city.

It was in 1661 that the school moved to the building currently occupied by the Durham University Music School to the north west of Palace Green.

Palace Green outside Durham Cathedral showing the location (left side of image) of Durham School from 1661 to 1844.

> There was some zeal for education in Durham during the 18th century. Durham School, rebuilt in 1661, on the Palace Green, soon became, instead of a local grammar school, a north-country public school of repute and wide influence. We can trace from the Restoration onwards not only the familiar city names such as Salvin, Wilkinson, Hutchinson, Blakiston, Fawcett, Bowes, Calverley, Cole. One of the chief distinctions of the school is the succession of local historians and antiquaries who drew their inspiration from the venerable association of the old school on the Green. Most famous of these is James Mickleton (1638–93), without whom no history of mediaeval or 17th-century Durham would be possible. Local history owes very much to Elias Smith, a notable head master (1640–66) who did his best to preserve the cathedral library through the Protectorate troubles, and to Thomas Rudd, head master (1691–9 and 1709–11), who indexed the Cathedral manuscripts. Later than these comes Thomas Randall (head master 1761–8), who made a large collection of manuscript material for local history books.
>
> —William Page, The city of Durham - Introduction (3 of 3) 1928.[4]

1844 to date

From its location on Palace Green outside Durham Cathedral, whilst Edward Elder was Headmaster[5] the school moved to its present site in 1844.[6]

The School has been steadily expanded and updated since then. For example:

Durham Cathedral seen from outside the chapel

> Henry Holden, Headmaster 1853 to 1882,[7] instigated new classrooms, dormitories, kitchens, sickroom then a sanatorium, bell tower and library.[]
>
> William Fearon, Headmaster 1882 to 1884, introduced the three term system used today and enlarged the playing fields and built an open air swimming pool.[] He was also responsible, in 1882, for the concrete path alongside the River Wear between Hatfield College Boat House and Elvet Bridge.[8] [9]
>
> Richard Budworth, Headmaster 1907 to 1932, oversaw great expansion in the School. In his tenure were added fives courts, a new boarding house in the Grove (a building now used for administration), Langley House, the Chapel (referred to below), new playing fields and classrooms, Poole House, the indoor swimming pool,[] the armoury and rifle range and the Kerr Arch.[] This arch, Grade II listed,[] was erected as the front entrance to the School and is in memory of G.C. Kerr, Cambridge Rowing Blue, Scottish rugby international and first civilian governor of the Sudan.[10]

The front of Durham School taken from the road outside, illustrating the Kerr Arch in the centre.

John Brett, Headmaster 1958 to 1967, saw the building of the cricket pavillion, new kitchens and laboratories and a new sanatorium.[]

Michael Vallance, Headmaster 1972 to 1982, opened a Junior House of boys age 11 to 13 named Ferens House (this closed in 2003), but the main developments in his tenure were the Budworth sports centre and Luce Theatre.[]

The key development in Michael Lang's time, Headmaster 1982, was the introduction of girls to the Sixth Form. There was also a new classroom block near the Budworth centre.[]

With Neil Kern as Headmaster, the principal developments included the introduction of girls throughout the School, making it fully co-educational, followed later by the creation of MacLeod House, and the all weather sports pitch near the Chapel.[11]

Ths school has been co-educational since 1985[12] and became independent from the Dean and Chapter of Durham Cathedral in 1996.[13]

School site

The school is located to the west of Durham Cathedral and across the River Wear from it. The campus consists of a range of buildings, some of them listed,[] and sports fields. The school has a boathouse located on the bank of the river, just downstream of Prebends Bridge.

Chapel and war memorial

The chapel from the main school buildings

The school chapel, built from 1924 to 1926 during the tenure of Richard Budworth as headmaster,[14] sits on top of a hill overlooking the main school site.

The building is used for services four times a week, with the major service held on Friday evenings, that the majority of pupils attend. The School is within the Anglican tradition.

The chapel is also the war memorial, its walls are engraved with the names of those who died in World War I and the further 79[15] who died in World War II. There are 97 steps to the chapel, one for each of the Old Dunelmians who died in WWI.[16] (Note that the UKNIWM reference and Durham School's disagree on the number of WWI dead - 97 or 98.) The steps were re-laid in 1954.[]

A further war memorial, which predates the building of the Chapel, can be seen in St Margaret's, the local Parish Church. It consists of a large brass panel listing the names of former pupils fallen in the First World War and is fixed on the south side of the Chancel arch, with the Parish memorial facing it on the north side.

The chapel steps

The chapel

Academic results and subjects

Subjects

A/AS Level

The School offers a range of subjects at A/AS Level, this varies from year to year as the needs of the students and the School change. Most students choose up to five subjects at AS Level, dropping to three at A2 Level.[]

GCSE

Durham School, in common with most other schools, offers a varied curriculum at GCSE where students study the obligatory mathematics, English and science (either the individual physics, chemistry and biology or a double award science course) and French. There are then a number of options that the students can choose from. At this time these do not include vocational GCSEs etc.[] The School use: the Camebrige IGCSE for English; EDEXCEL's maths GCSE for maths; and both the aqa duel science GCSE and the AQA iGCSE separate science GCSE.

Key Stage 3 (Years 7 to 9)

Durham School offers a range of subjects to students in Years 7 to 9 that lead into the GCSE programme starting, generally, in Year 10, with the exception of those in set 1 maths who do there maths GCSE a year early. They start their course in year 9.

Results

Durham School students cover a wide range of abilities from the academically gifted to those whose talents lie in other areas,[17] this is reflected in the results that the School achieves as a whole. The School states that "We aim to bring out the best in every girl and boy and work together with our pupils to ensure that the grades they achieve in their public examinations are a true reflection of their full potential."[18]

Pastoral care

Pastoral care at Durham School is based around the house system:

House system

Although lessons are co-educational and sport takes place generally in year groups, for pastoral and sporting competition purposes, the school is divided into five different single sex houses.[19] Each student is assigned to a house at the start of his or her time in the school and will, usually, remain in that house for the whole of their school career. The houses are separate buildings each with its own character in which the students are able to use the facilities and do private study. Currently the houses comprise:

House	Occupants	Founded	House colour
The Caffinites	boys, day, boarding (Sixth form pupils only)	1847[]	
MacLeod	girls, day	2005[20]	
Pimlico	girls, boarding and day	1985[21]	
Poole	boys, boarding and day	1924[22]	
The School House	boys, boarding and day	1844[23]	

School House boys are often nicknamed Bungites after the Headmaster Henry Holden who was also their Housemaster and himself nicknamed 'Bung' due to his ability to tell tall tales.[]

Caffinites was called The Second Masters House until it was renamed in 1924.[]

Co-curricular

Sport

With facilities that encompass playing fields, swimming pool, all weather pitch and boathouse, Durham School has a wide ranging sports programme that includes athletics, badminton, cricket, cross country, fencing, football, gym, hockey, netball, rounders, rowing, rugby, squash, swimming, tennis golf' and water polo.[24]

Cricket

Along with rowing, cricket is the major boys sport for the summer term. The school considers cricket to be one of its high profile sports with the 1st XI competing against a number of high profile teams such as the MCC.[25] [26] The cricket pavilion was built in 1960.[]

Hockey

The main sport for girls in winter is hockey. Each year a number of girls represent the county. Hockey is also played by the boys, predominately in the Easter term.[27]

The all weather pitch

Rowing

Rowing has been a major sport at Durham School since at least 1838,[28] and today it continues to flourish. The Club aims to provide rowing opportunities to any who wish to be involved - "All year groups at the school have the opportunity to row at the appropriate level for their physiological and physical development"[29] - and also wins at regattas and head races at home and abroad. In addition, the Club offers rowing to students, parents, friends and other local schools.

Rugby

The School has a number of playing fields, used predominately for Rugby. The largest of these, Top Ground, was acquired in 1918.[] This has required work on its drainage over the years, 1921 and 1925[] and again in 2009.

Durham School is the fourth or fifth oldest football club of any kind in the world[30] [] and has produced many international rugby union players as well as introducing the game to Scotland.[] Durham School Football Club was founded in 1850[31] and was from 1890-1930 one of the rugby nurseries in England - this was first recognised by the Barbarians who honoured the school by playing a full side against the Old Dunelmians in 1897 (Barbarians won 18-5). Just before the First World War the school supplied four England players as well as a number of England trialists. Between 1920 and 1930 the school produced seven full internationals as well as several trialists (England, Scotland and British Lions). The headmaster during this great period was Richard Thomas Dutton Budworth who was himself a former England international and Barbarian.[32] The history of the sport is celebrated in the annual Veterrimi IV Rugby Tournament hosted by Durham School and consisting of Rugby School, Cheltenham College and Sherborne School. In the inaugural event, October 2010, Rugby School won the final, beating Durham School 10-7.[33]

Newcastle Falcons (originally Gosforth) rugby union club were founded as the old boys side and played in the school's colours until recently (green and white hoops).[34]

Combined Cadet Force

The Combined Cadet Force (CCF) is a Ministry of Defence sponsored youth organisation within schools in Great Britain.

Durham School CCF has been running since it was formed as the Officer Training Corps (OTC) in 1914.[35] The CCF is based in the Armoury and Range (previously there had been an air rifle range set up in the School House lavatories in 1906[36]) built in 1926 and 1929 respectively.[]

Currently the CCF is a voluntary activity that students may opt into from Year 9 onwards.

There are three sections in the school's CCF:[37]

- The Army section
- The Royal Navy section
- The Royal Air Force section

Music

Centred on the Grade II listed Music School,[] Music is an important aspect of life at Durham School with a high profile in the School itself and also known in the region as a whole with concerts at The Sage Gateshead and in Durham Cathedral from time to time. The School offers music as an academic subject at both GCSE and A Level.[38]

Drama

Centred on the Luce Theatre, also Grade II listed,[] a versatile space named after Henry Luce, Headmaster 1932 to 1958, drama at Durham School is a key part of the academic and extra-curricular provision. Productions are varied and take place two or three times per year for different age groups.[39]

Bow, Durham School

Bow, Durham School is the Durham School preparatory institution for pupils between the ages of 3 to 11 years. Founded in 1885 as **Bow School**, it was an all-male institution until becoming coeducational in 2006.[] The campus is situated one-half mile to the east of the senior school, overlooking the World Heritage Site, Durham Cathedral.[] Former pupils are known as "Old Bowites" and several prominent alums include British politicians and nobles, such as William Fletcher-Vane, 1st Baron Inglewood, Sir Brian Horrocks and Sir Gilbert Longden, and a number of professional sportsmen, such as Michael Philip Weston.[40]

Notable past pupils: Old Dunelmians

The past pupils of Durham School, referred to as Old Dunelmians, have been, and continue to be, found across the spectrum of public life, the armed services, the arts, the church and in sport. "Dunelmian" is derived from Dunelmensis, the Latin adjective of Durham. Their number include Granville Sharp,[41] founder of Sierra Leone; Field Marshal Henry Hardinge, 1st Viscount Hardinge,[42] commander in chief of the British Army after The Duke of Wellington; Anthony Salvin,[43] the architect of national fame, Michael Gough,[44] the actor best known for playing "Alfred" in the first four of the modern Batman films and comedian Alexander Armstrong.[45]

See also

- Durham School Boat Club
- Headmasters of Durham School
- List of Old Dunelmians

References

[1] http://www.durhamschool.co.uk/old-dunelmiansdistinguished.asp
[40] http://www.durhamschool.co.uk/old-bowites.asp

External links

- School website (http://www.durhamschool.co.uk/)
- Bow, Durham School (http://www.durhamschool.co.uk/prep-school.asp)
- Boat Club website (http://dsbc.synthasite.com/)
- Old Dunelmians website (http://www.olddunelmian.com/)
- Independent Schools Inspectorate Inspection Reports (http://www.isi.net/schools/6415/)

University_of_Bristol

The **University of Bristol** is a research university located in Bristol, United Kingdom.[] It received its Royal Charter in 1909,[] and its predecessor institution, University College, Bristol, had been in existence since 1876.[]

Bristol has been named amongst the world's top 30 universities by the QS World University Rankings.[] A highly selective institution, it has an average of 14 applicants for each undergraduate place. The University had a total income of £426.7 million in 2011/12, of which £112 million was from research grants and contracts.[] It is the largest independent employer in Bristol.[1]

Current academics include 18 Fellows of the Academy of Medical Sciences, 10 Fellows of the British Academy, 13 Fellows of the Royal Academy of Engineering and 31 Fellows of the Royal Society.[2]

Bristol is a member of the Russell Group of research-intensive British universities,[3] the European-wide Coimbra Group[] and the Worldwide Universities Network, of which the University's Vice-Chancellor Prof. Eric Thomas was Chairman from 2005 to 2007.[]

History

Foundation

The earliest antecedent of the university was the engineering department of the Merchant Venturers' Technical College (founded as a school as early as 1595) which became the Engineering faculty of Bristol University.[4] The University was also preceded by Bristol Medical School (1833) and University College, Bristol, founded in 1876,[] where its first lecture was attended by only 99 students.[5] The University was able to apply for a Royal Charter due to the financial support of the Wills and Fry families, who made their fortunes in tobacco plantations and chocolate, respectively. The Wills Family made a vast fortune from the tobacco industry and gave generously to the city and University. The Royal Charter was gained in May 1909, with 288 undergraduates and 400 other students entering the University in October 1909. Henry Overton Wills III became its first chancellor.[] The University College was the first such institution in the country to admit women on the same basis as men.[] However, women were forbidden to take examinations in medicine until 1906.[]

Historical development

There shall be from henceforth for ever in Our said City of Bristol a University...
King Edward VII, Charter of Incorporation of the University of Bristol, 4 December 1909[]

Since the founding of the University itself in 1909, it has grown considerably and is now one of the largest employers in the local area, although it is smaller by student numbers than the nearby University of the West of England.[] Bristol does not have a campus but is spread over a considerable geographic area. Most of its activities, however, are concentrated in the area of the city centre, referred to as the "University Precinct". It is a member of the Russell Group of research-led UK universities, the Coimbra Group of leading European universities and the Worldwide Universities Network (WUN).

Early years

After the founding of the University College in 1876, Government support began in 1889. After mergers with the Bristol Medical School in 1893 and the Merchant Venturers' Technical College in 1909,[6] this funding allowed the opening of a new Medical School and an Engineering School—two subjects that remain among the University's greatest strengths. In 1908, gifts from the Fry and Wills families, particularly £100,000 from Henry Overton Wills III (£6m in today's money), were provided to endow a University for Bristol and the West of England, provided that a Royal Charter could be obtained within two years. In December, 1909, the King granted such a Charter and erected the University of Bristol.[] Henry Wills became its first Chancellor and Conwy Lloyd Morgan the first Vice-Chancellor.[] Wills died in 1911 and in tribute his sons George and Harry built the Wills Memorial Building, starting in 1913 and finally finishing in 1925.[7] Today, it houses parts of the academic provision for earth sciences and law, and graduation ceremonies are held in its Great Hall. The Wills Memorial Building is a Grade II* listed building.[8]

The Wills Memorial Building (Schools of Law and Earth Sciences) on Park Street, Bristol. The tower was cleaned in 2006–2007.

In 1920, George Wills bought the Victoria Rooms and endowed them to the University as a Students' Union.[] The building now houses the Department of Music and is a Grade II* listed building.[9]

King's College London students at the University of Bristol in 1940

At the point of foundation, the University was required to provide for the local community. This mission was behind the creation of the Department of Extra-Mural Adult Education in 1924 to provide courses to the local community. This mission continues today; a new admissions policy specifically caters to the 'BS' postcode area of Bristol.[]

Among the famous names associated with Bristol in this early period is Paul Dirac, who graduated in 1921 with a degree in engineering, before obtaining a second degree in mathematics in 1923 from Cambridge. For his subsequent pioneering work on quantum mechanics, he was awarded the 1933 Nobel Prize for Physics.[10] Later in the 1920s, the H.H. Wills Physics Laboratory was opened by Ernest Rutherford.[11] It has since housed several Nobel Prize winners: Cecil Frank Powell (1950);[12] Hans Albrecht Bethe (1967);[13] and Sir Nevill Francis Mott (1977).[14] The Laboratory stands on the same site today, close to the Bristol Grammar School and the city museum.

Sir Winston Churchill became the University's third Chancellor in 1929, serving the University in that capacity until 1965.[] He succeeded Richard Haldane who had held the office from 1912 following the death of Henry Wills.[] []

During World War II, the Wills Memorial was bombed, destroying the Great Hall and the organ it housed.[] It has since been restored to its former glory, complete with oak panelled walls and a new organ.

Post-war development

In 1946, the University established the first drama department in the country.[] In the same year, Bristol began offering special entrance exams and grants to aid the resettlement of servicemen returning home. Student numbers continued to increase, and the Faculty of Engineering eventually needed the new premises that were to become Queen's Building in 1955. This substantial building housed all of the University's engineers until 1996, when Electrical Engineering and Computer Science moved over the road into the new Merchant Venturers' Building to make space for these rapidly expanding fields. Today, Queen's Building caters for most of the teaching needs of the Faculty and provides academic space for the "heavy" engineering subjects (civil, mechanical, and aeronautical).

With unprecedented growth in the 1960s, particularly in undergraduate numbers, the Student's Union eventually acquired larger premises in a new building in the Clifton area of the city, in 1965. This building was more spacious than the Victoria Rooms, which were now given over to the Department of Music. The new Union provides many practice and performance rooms, some specialist rooms, as well as three bars: Bar 100; the Mandela (also known as AR2) and the Avon Gorge. Whilst spacious, the Union building is thought by many to be ugly[15] and out of character compared to the architecture of the rest of the Clifton area, having been mentioned in a BBC poll to find the worst architectural eyesores in Britain.[16] The University has proposed relocating the Union to a more central location as part of its development 'masterplan'.[17] More recently, plans for redevelopment of the current building have been proposed.[18]

The 1960s were a time of considerable student activism in the United Kingdom, and Bristol was no exception. In 1968, many students marched in support of the Anderson Report, which called for higher student grants. This discontent culminated in an 11-day sit-in at the Senate House (the administrative headquarters of the University).[] A series of Chancellors and Vice-Chancellors led the University through these decades, with Henry Somerset, 10th Duke of Beaufort taking over from Churchill as Chancellor in 1965 before being succeeded by Dorothy Hodgkin in 1970 who spent the next 18 years in the office.[]

As the age of mass higher education dawned, Bristol continued to build its student numbers. The various undergraduate residences were repeatedly expanded and, more recently, some postgraduate residences have been constructed. These more recent ventures have been funded (and are run) by external companies in agreement with the University.

Since 1988, there have been only two further Chancellors: Sir Jeremy Morse, then chairman of Lloyds Bank who handed over in 2003 to Brenda Hale, the first female Law Lord.[] []

The Victoria Rooms. Comprising the world renowned School of Music.

One of the few Centres for Deaf Studies in the United Kingdom was established in Bristol in 1981, followed in 1988 by the Norah Fry Centre for research into learning difficulties. Also in 1988, and again in 2004,[19] the Students' Union AGM voted to disaffiliate from the National Union of Students (NUS). On both occasions, however, the subsequent referendum of all students reversed that decision and Bristol remains affiliated to the NUS.

In 2002, the University was involved in argument over press intrusion after details of the son of then-Prime Minister Tony Blair's application to university were published in national newspapers.

As the number of postgraduate students has grown (particularly the numbers pursuing taught Master's Degrees), there eventually became a need for separate representation on University bodies and the Postgraduate Union (PGU) was established in 2000.[20] Universities are increasingly expected to exploit the intellectual property generated by their research activities and, in 2000, Bristol established the Research and Enterprise Division (RED) to further this cause (particularly for technology-based businesses). In 2001, the university signed a 25-year research funding deal with IP2IPO, an intellectual property commercialisation company.[21] In 2007, research activities were expanded further with the opening of the Advanced Composites Centre for Innovation and Science (ACCIS) and The Bristol Institute for Public Affairs (BIPA).

In 2002, the University opened a new Centre for Sports, Exercise and Health in the heart of the University precinct.[22] At a cost, local residents are also able to use the facilities.[23]

Most of the buildings here are used by the University. The Wills Memorial Building is left of centre. Viewed from the Cabot Tower on Brandon Hill

Expansion of teaching and research activities continues. In 2004, the Faculty of Engineering completed work on the Bristol Laboratory for Advanced Dynamics Engineering (BLADE). This £18.5m project[24] provides cutting-edge technology to further the study of dynamics and is the most advanced such facility in Europe. It was built as an extension to the Queen's Building and was officially opened by Queen Elizabeth II in March 2005.

In January, 2005, The School of Chemistry was awarded £4.5m by the Higher Education Funding Council for England to create Bristol ChemLabS: a Centre for Excellence in Teaching & Learning (CETL),[25] with an additional £350k announced for the capital part of the project in February, 2006. Bristol ChemLabS stands for Bristol Chemical Laboratory Sciences; it is the only Chemistry CETL in the UK.

September 2009 saw the opening of the University's Centre for Nanoscience and Quantum Information. This £11 million state of the art building is dubbed as the quietest building in the world and has other technologically sophisticated features such as self-cleaning glass. Advanced research into quantum computing, nanotechnology, materials and other disciplines are being undertaken in the building.[26]

There is also a plan to significantly redevelop the centre of the University Precinct in the coming years.[27] The first step began in September 2011, with the start of construction of a state-of-the-art Life Sciences building. In a time of heavy financial pressures on all Universities, this £50 million project is a clear statement that Bristol is committed to world class research and teaching facilities.[28]

2003 admissions controversy

The University has been regarded as being elitist by some commentators,[29] taking 41% of its undergraduate students from non-state schools, according to the most recent 2009/2010 figures, despite the fact that such pupils make up just 7% of the population[30] and 18% of 16+ year old pupils across the UK.[31] The high ratio of undergraduates from non-state school has led to some tension at the university.[32] In late February and early March 2003, Bristol became embroiled in a row about admissions policies, with some private schools threatening a boycott[33] based on their claims that, in an effort to improve equality of access, the University was discriminating against their students. These claims were hotly denied by the University.[34] In August, 2005, following a large-scale survey, the Independent Schools Council publicly acknowledged that there was no evidence of bias against applicants from the schools it represented.[35] The University has a new admissions policy,[] which lays out in considerable detail the basis on which any greater or lesser weight may be given to particular parts of an applicant's backgrounds—in particular, what account may be taken of which school the applicant hails from. This new policy also encourages greater participation from locally resident applicants.

Campus

The Great Hall of the Wills Memorial Building, here used for an award ceremony for the Queen Elizabeth's Hospital.

Some of the University of Bristol's buildings date to its pre-charter days when it was University College Bristol. These buildings were designed by Charles Hansom, the younger brother of Joseph Hansom, Joseph being the inventor of the Hansom Cab. These buildings suffered being built in stages due to financial pressure. George Oatley added to them a tower in memory of Albert Fry which can still be seen on University Road. The first large scale building project the University of Bristol undertook on gaining a charter was the Wills Memorial Building which it was hoped would be a symbol of academic permanence for the University and a memorial to the chief benefactor of the University Henry Overton Wills. It was requested to the architect George Oatley that the building be built to last at least 400 years but the site purchased, at the top of Park Street suffered from an awkward slope and a desirability to link the building with the Museum and Art Gallery situated adjacent to the plot. The architecture critic Roger Gill has stated that the building is "remarkable in size" but noted that the "ambience of a medieval University was strangely lacking". He goes on to criticise the building as a "sham" and a "folly".[36] The armorials on the Founder's Window represent all of the interests present at the founding of the University of Bristol including the Wills and Fry families. The Tyndalls Park Estate and Royal Fort House were also purchased from the trustees of the Tyndall family allowing the University to expand. Many Departments in the Faculty of Arts are housed in large Victorian houses which have been converted for teaching.[37]

Goldney gardens entered the property of the University of Bristol through George Wills who had hoped to build an all male hall of residence there. This was prevented due to the moral objection of the then warden of Clifton Hall House who objected to the idea of male and female residences being in such close proximity. University records show that Miss Starvey was prepared to resign over the issue and that she had the support of the then Chancellor Conwy Lloyd Morgan.[38] Eventually land was purchased in Stoke Bishop allowing Wills Hall to be bought, allowing the building of what has been described as a "quasi-Oxbridge" hall, to which was added the Dame Monica Wills Chapel added by George Wills' widow after his death.

The Gardens of Goldney Hall were acquired by the Wills family

Burwalls, a mansion house on the other side of the Avon Gorge, was used as a halls of residence in the past and was a home of Sir George Oatley. The building is now used to house the Centre for Continuing Education.[]

Many of the more modern buildings, including Senate House and the newer parts of the HH Wills Physics Laboratory, were designed by Raplh Brentnall after funds from the University Grants Committee. He is also responsible for the extension to the Wills Memorial Building library which was completed to such standard that few now realise that is an extension to the original building.[39] Brentnall oversaw the rebuilding of the Great Hall of the Wills Memorial Building after it was partly destroyed during the Bristol Blitz of World War II. The buildings of St Michael's Hill were rebuilt using hundreds of old photographs in order to recreate the original houses. The flats at Goldney Hall were designed by Michael Grice and received an award from the Civic Trust for their design.[40] Bristol University owns some of the best examples of Georgian architecture in the city, the best examples being Royal Fort House, Clifton Hill House and Goldney Hall despite some additions.[41] The Victoria Rooms which house the Music Department were designed by Charles Dyer and is seen as a good example of a Greek revival movement in British architecture. The tympanum of the building depicts a scene from *The Advent of Morning* designed by Jabez Tyley.[42] Its major feature was a large organ which has since been destroyed by fire.

Academic reputation

ARWU[] **(2012, national)**	8
ARWU[] **(2012, world)**	70
QS[] **(2012/13, world)**	28
THE[] **(2012/13, national)**	9
THE[] **(2012/13, world)**	66
Complete[] **(2014, national)**	15
The Guardian[] **(2014, national)**	18
The Sunday Times[43] **(2013, national)**	10
The Times[44] **(2013, national)**	11

The Geography department

Bristol is known for academics, excellent facilities, and a desirable location. League tables usually place Bristol within the top ten universities in the United Kingdom and it attracts many academically gifted students. For example, the 21 July 2011 edition of Times Higher Education reported that Bristol was fifth in a UK league table for the highest proportion of students with A-level grades AAB or better. Internationally, the 2012/2013 QS World University Rankings[45] placed Bristol at 28th overall in the world. The rankings also placed Bristol at 15th in the world in terms of reputation with employers, placing higher than several American Ivy League universities, including Princeton University, Cornell and UPenn.[46] Another international ranking, the Shanghai Jiao Tong University Academic Ranking of World Universities, placed Bristol 70th globally in 2011[] The Times Higher Education World University Ranking placed Bristol at 66th in 2011.[47] According to data published in *The Telegraph* Bristol has the third-highest percentage of 'good honours' of any UK

Royal Fort and the Physics department

university, behind Oxford and Cambridge.[48] In the 2010 Centre for Higher Education's Development's Excellence Rankings, Bristol is one of only four UK Universities (Oxford, UCL and Manchester) to be rated Excellent in all seven departments.[49] The following courses offered by University of Bristol, managed to reach top 5 in the Times ranking (2008): Computer Science(3rd); Electrical and Electronic Engineering(3rd); Civil Engineering(5th); Biological Sciences(3rd); Mathematics (3rd); and Psychology (4th). Furthermore, the QS World University Rankings place Bristol in the world's top 100 universities for all subject areas in 2011: Arts and Humanities (57th), Natural Sciences (40th), Engineering & IT (83rd), Social Sciences (65th) and Life Sciences (70th).[50] A further breakdown of the QS World University Natural Sciences Ranking shows the following: Earth Sciences (25th),[51] Mathematics (35th),[52] Environmental Sciences (39th),[53] Physics (41st),[54] and Chemistry (48th).[55]

In addition, Bristol is particularly strong in the field of social sciences, particularly in Economics, Finance and Management, and was recently rated 4th in the 2008 Guardian University Guide for Business and Management Studies.[56]

In 2011, The Guardian also ranked Bristol as 3rd in the UK for Geography, just behind 2nd place Oxford[57] and ranked Bristol as 1st in the UK for Music.[58]

In The Complete University Guide 2013, Bristol ranked 5th for German,[59] 4th for Russian,[60] 3rd for Mechanical and Civil Engineering,[61] 3rd for Music[62] and 2nd for Drama.[63] Bristol is also known for its research strength, having 15 departments gaining the top grade of 5* in the 2001 Research Assessment Exercise. Overall, 36 out of 46 departments rated gained the top two ratings of 5 or 5*, and 76% of all the academic staff working in departments scored these top two levels.[64] [65] In terms of teaching strength, Bristol had an average Teaching Quality Assessment score of 22.05/24 before the TQA was abolished.[66] For admission in October 2010, Bristol reported an average of 10.2 applications per place with the average A-level score on admission being 478.5.[67] Bristol's drop-out rate is also lower than the benchmark set by HEFCE of no more than 3.1%.[68]

Academic structure

The University is made up of a number of schools and departments organised into six faculties:[69]

Faculty of Arts

The Wills Memorial Library of Earth Sciences

- Archaeology and Anthropology
- Drama: Theatre, Film and Television
- History of Art
- Music
- Philosophy
- Classics and Ancient History
- English
- History
- Religion and Theology
- French
- German
- Hispanic, Portuguese and Latin American Studies
- Italian
- Russian (which also includes Czech)

Faculty of Engineering

- Aerospace Engineering
- Civil Engineering
- Computer Science
- Electrical and Electronic Engineering
- Engineering Mathematics
- Mechanical Engineering

Faculty of Medical and Veterinary Sciences

The Faculty of Engineering Queen's Building

- Comparative and Clinical Anatomy
- Biochemistry
- Cellular and Molecular Medicine
- Veterinary Sciences
- Physiology and Pharmacology

Faculty of Science

- Biological Sciences
- Chemistry
- Earth Sciences
- Experimental Psychology
- Geographical Sciences
- Mathematics
- Physics

The School of Chemistry

Faculty of Medicine and Dentistry

- Clinical Sciences
- Social and Community Medicine
- Oral and Dental Sciences

Faculty of Social Sciences and Law

- Social Work
- Education (Graduate School of)
- Geographical Sciences *(affiliated)*
- Policy Studies
- Deaf Studies
- Economics, Finance and Management (EFM)
- Law
- Sociology Politics and International Studies (SPAIS)

Degrees

Bristol awards a range of academic degrees spanning bachelor's and master's degrees as well as junior doctorates and higher doctorates. The postnominals awarded are the degree abbreviations used commonly among British universities. The University is part of the Engineering Doctorate scheme,[70] and awards the Eng. D. in systems engineering, engineering management, aerospace engineering and non-destructive evaluation.[71]

Bristol notably does not award by title any Bachelor's degrees in music, which is available for study but awarded B.A. (although it does award M.Mus. and D.Mus.), nor any degree in divinity, since divinity is not available for study (students of theology are awarded a B.A.). Similarly, the University does not award B.Litt. (Bachelor of Letters), although it does award both M.Litt. and D.Litt. In regulations, the University does not name M.D. or D.D.S. as higher doctorates, although they are in many universities.[72] as these degrees are normally accredited professional doctorates.

The degrees of D.Litt., D.Sc., D.Eng., LL.D. and D.Mus., whilst having regulations specifying the grounds for award,[73] are most often conferred as honorary degrees (*in honoris causa*).[74] Those used most commonly are the D.Litt., D.Sc. and LL.D., with the M.A. (and occasionally the M.Litt.) also sometimes conferred honorarily for distinction in the local area or within the University.

Governance

In common with most UK universities, Bristol is headed formally by the Chancellor, currently Baroness Hale of Richmond [75] and led on a day-to-day basis by the Vice-Chancellor, currently Prof Eric Thomas, who is the academic leader and chief executive. There are two Pro Vice-Chancellors [76] and three ceremonial Pro-Chancellors.[] The Chancellor may hold office for up to ten years and the Pro-Chancellors for up to three, unless the University Court determines otherwise,[77] [78] but the Vice-Chancellor and Pro-Vice-Chancellors have no term limits.[79] [80] The Vice Chancellor is supported by a Deputy Vice-Chancellor [81].

Responsibility for running the University is held at an executive level by the Vice-Chancellor, but the Council is the only body that can recommend changes to the University's statutes and Charter,[] with the exception of academic ordinances. These can only be made with the consent of the Senate, the chief academic body in the University which also holds responsibility for teaching and learning, examinations and research and enterprise.[] [82] The Chancellor and Pro Chancellors are nominated by Council and appointed formally by Court, whose additional powers are now limited to these appointments and a few others, including some lay members of Council.[83] Finally, Convocation, the body of all staff, ceremonial officers and graduates of the University, returns 100 members to Court and one member to Council,[] but is otherwise principally a forum for discussion and to ensure graduates stay in touch with the University.

Student life

Students' Union

The University of Bristol Union building

The University of Bristol Union (UBU) located on Queen's Road is a founding member of the National Union of Students and is amongst the oldest students' unions in England. The Union oversees the two media outlets of the University, the radio station *BURST* (Bristol University Radio Station) and the student newspaper *Epigram*. In terms of student life, the Union is responsible for the organisation of the annual freshers' fair, the coordination of Bristol Student Community Action, which organises volunteering projects in the local community, and the organisation of entertainment events and student societies. **Bristol Improv** are a society which regularly performs improvisational comedy for students and locals alike in a number of free shows every month.[84] Previous presidents have included Sue Lawley and former Liberal Democrat MP Lembit Öpik. There is a separate union for postgraduate students, as well as an athletic union, which is a member of the British Universities & Colleges Sport.[85] In distinction to the 'blues' awarded for sporting excellence at Oxford and Cambridge, Bristol's outstanding athletes are awarded 'reds'.[86]

Halls of residence

Accommodation for students is primarily in the central precinct of the University and two areas of Bristol: Clifton and Stoke Bishop.[] In Stoke Bishop, Wills Hall on the edge of the Clifton Downs was the first to be opened, in 1929, by then-Chancellor Winston Churchill. Its original quadrangle layout has been expanded twice, in 1962 and 1990.[] Churchill Hall, named for the Chancellor, followed in 1956, then Badock Hall in 1964.[] [87] At the time of Badock Hall's establishment, some of the buildings were called Hiatt Baker Hall, but two years later, Hiatt Baker moved to its own site and is now the largest hall in the University.[] [88] The first self-catering hall in Stoke Bishop was University Hall, established in 1971 with expansion in 1992.[] The University's newest undergraduate residence, Durdham Hall, was opened in Stoke Bishop in 1994.[] All of the main halls elect groups of students to the Junior Common Room to organise the halls social calendar for the next year.

Wills Hall

Goldney Hall

In Clifton, Goldney Hall was built first in the early 18th century by a wealthy merchant family of the same surname and eventually became part of the University in 1956.[89] It is a popular location for filming, with *The Chronicles of Narnia*, *The House of Eliott* and *Truly, Madly, Deeply*, as well as episodes of *Only Fools and Horses* and *Casualty*, being filmed there.[90] The Grotto in the grounds is a Grade I listed building.[91] Clifton Hill House is another Grade I listed building now used as student accommodation in Clifton. The original building was constructed between 1745 and 1750 by Isaac Ware, and has been used by the University since its earliest days in 1909.[] [92] Manor Hall comprises five separate buildings, the principal of which was erected from 1927–1932 to the design of George Oatley following a donation from Henry Herbert Wills.

One of its annexes, Manor House, has recently been refurbished and officially 'reopened' in 1999.[] [93] Goldney Hall has beautiful gardens and modern accommodation complexes. Clifton Hill House has more dated facilities, but as with all the Clifton residences also possesses attractive gardens. Manor Hall houses the largest and most dated rooms, some dating back to the early 20th century. The hall's gardens are breathtakingly attractive, and complement what is a historically beautiful hall.

Clifton Hill House

On the central precinct sits The Hawthorns, a student house accommodating 115 undergraduate students.[94] The house started life as a collection of villas built somewhere between 1888 and 1924[95] that were later converted, bit by bit, into a hotel by John Dingle.[96] The Hawthorns also houses conferencing facilities, the staff refectory and bar, the Accommodation Office and the Student Houses Office. Several of the residences in the central precinct are more recent and have been built and are managed by third-party organisations under exclusivity arrangements with the University. These include Unite House and Chantry Court, opened in 2000 and 2003 respectively by the UNITE Group,[97] [98] [99] as well as Dean's Court (2001, postgraduates only) and Woodland Court (2005), both run by the Dominion Housing Group.[100] [101]

Symbols

In common with other universities in the United Kingdom, Bristol uses its particular pattern of academic dress as well its logo and coat of arms to represent itself.

The University coat of arms

Academic dress

The University specifies a mix of Cambridge and Oxford academic dress. For the most part, it uses Oxford-style gowns and Cambridge-style hoods, which are required to be 'University red'[] (see the logo at the top of the page)

Logo and arms

In 2004, the University unveiled its new logo. The icons in the logo are the sun for the Wills family, the dolphin for Colston, the horse for Fry and the ship-and-castle from the mediaeval seal of the City of Bristol, as also used in the coat of arms. The shape of the whole logo represents the open book of learning.[] This logo has replaced the University arms shown, but the arms continue to be used where there is a specific historical or ceremonial requirement. The arms comprise:

> *argent on a cross quadrate gules the arms of the City of Bristol between in pale and a sun in splendour (for Wills) and an open book proper, leaved and clasped or, and inscribed with the words Nisi quia Dominus, and in fesse to the sinister a dolphin embowed (for Colston), and to the dexter a horse courant (for Fry), both of the third.*

The inscription on the book is the Latin opening of the 124th Psalm, *"If the Lord Himself had not (been on our side...)"*.[]

Notable people

Bristol is associated with 11 Nobel Laureates, and current academics include 18 Fellows of the Academy of Medical Sciences, 10 Fellows of the British Academy, 13 Fellows of the Royal Academy of Engineering and 31 Fellows of the Royal Society.[102]

- Sir Michael Berry, knighted in 1996, one of the discoverers of quantum mechanics' 'geometric phase'[103]
- John Rarity who, in 2001, set a then world-record 1.9 km range for free-space secure key exchange using quantum cryptography[104]
- David May, founder of XMOS and lead architect for the transputer
- Mark Horton, a British maritime and historical archaeologist and one of the presenters of the BBC's *Coast* television series[105]

Patricia Broadfoot, Vice-Chancellor of the University of Gloucestershire, and Nigel Thrift, Vice-Chancellor of the University of Warwick both were previously faculty at Bristol.[106] [107] Anthony Epstein, co-discoverer of the Epstein-Barr virus, was Professor of Pathology at the University from 1968–1982.[108] Historical academics include Sir John Lennard-Jones, discoverer of the Lennard-Jones potential in physics[109] [110] and Alfred Marshall, one of the University College's Principals and influential economist in the latter part of the 19th century.[111] Rohit Parikh lectured in the mathematics department from 1965 to 1967, as did Brian Rotman for twenty years.

University of Bristol is associated with 2 Ig Nobel Prizes. Sir Michael Berry shared the award (with Andre Geim, a Nobel Laureate) for using magnets to levitate a frog.[112] Professor Gareth Jones also shared an Ig Nobel prize for scientifically documenting fellatio in fruit bats.[113]

Alumni

Notable alumni of the University of Bristol include writers Dick King-Smith, Angela Carter and David Nicholls, author of the novel *Starter for Ten*, turned into a screenplay set in the University of Bristol.[114] Other high-profile former students include BBC News' Chief Political Correspondent James Landale (who founded the Bristol University independent newspaper, *Epigram*), Paul Dirac (Lucasian Professor at Cambridge and Nobel Laureate), editor-in-chief of the Telegraph Media Group William Lewis (journalist), illusionist Derren Brown, author of business books Mark Simmons (author), Global Economist], author, commentator and Executive Vice Chair of the Work Foundation Will Hutton, Serial award winning entrepreneur Mike Bennett (businessman) of digital agency E3 Media (digital agency)], Prince of Monaco Albert II, TV newsreader Alastair Stewart, the Sky News US Correspondent Dominic Waghorn, as well as musician James Blunt. Radio 4 presenter Sue Lawley was also a student there, whilst former Liberal Democrat MP Lembit Öpik was President of Bristol University Students' Union during his time there. Jack Harries, of YouTube fame for his channel JacksGap, attended the university for a semester in 2012, before dropping out to pursue a career in filmmaking with his brother Finn. He read Drama.

The University also has a comedy pedigree. *Little Britain* stars Matt Lucas and David Walliams,[115] attended the university, as did Simon Pegg (of *Hot Fuzz* fame), Chris Morris, creator of the controversial *Brass Eye* and Jon Richardson. Other comedy stars include Laura Crane-Brewer of The Office fame and Chris Langham, of The Thick of It fame, standup comic Marcus Brigstocke, and Radio 4 favourite Danny Robbins. More recently, Bristol students established a satirical newspaper, The Tart,[116] which received national press attention.[117]

Notable alumni from the Film and Television Production department include film directors Mick Jackson, Michael Winterbottom, Marc Evans, Christopher Smith, Alex Cox and Peter Webber amongst many others.

For a full list of famous alumni, see Bristol University's page on notable alumni [118].

See also

- CHOMBEC
- Education in Bristol
- Third oldest university in England debate
- List of modern universities in Europe (1801–1945)

Notes and references

[36] Carleton (1984), p129
[37] Higher Education Quality Council (1993), p2
[38] Carleton (1984), p132
[39] Carleton (1984), p138
[40] Carleton (1984), p139
[41] Carleton (1984), p141
[42] Carleton (1984), p136
[43] (subscription required)
[44] (subscription required)
[47] World University Rankings 2011-2012 - Times Higher Education (http://www.timeshighereducation.co.uk/world-university-rankings/2011-2012/top-400.html)
[75] http://www.bristol.ac.uk/university/chancellor.html
[76] http://www.bristol.ac.uk/university/management/pvc
[81] http://www.bristol.ac.uk/university/management/dvc/
[84] http://www.bristolimprov.co.uk Retrieved 6 December 2007.
[95] Bristol Record Office: Volume 19, Folio 74 and Volume 22, Folio 34
[112] Error: Bad DOI specified: 10.1088/0143-0807/18/4/012
[113] Error: Bad DOI specified: 10.1371/journal.pone.0007595
[117] Student news goes national – The Times (http://www.timesonline.co.uk/tol/life_and_style/education/student/how_to_get_a_job/article2812463.ece)

[118] http://www.bristol.ac.uk/alumni/featured/

Further reading

- Carleton, Don (1984) "University for Bristol: A History in Text and Pictures". University of Bristol P. ISBN 0-86292-200-3
- Delany, Rosalind (2002) "How Did This Garden Grow?: The History of the Botanic Gardens of the University of Bristol". Friends of Bristol University Botanic Garden ISBN 0-9543504-0-5
- Crossley Evans, M. J. (1994) "A History of Wills Hall University of Bristol". University of Bristol P. ISBN 0-86292-421-9
- Whittingham, Sarah (2003) "Wills Memorial Building". University of Bristol ISBN 0-86292-541-X

External links

- University of Bristol website (http://bristol.ac.uk/)
- University of Bristol Students' Union (http://www.ubu.org.uk/)

Video clips

- Computer Science Department YouTube channel (http://uk.youtube.com/user/CSBristol)

Jazz

Jazz is a music genre that originated at the beginning of the 20th century, arguably earlier, within the African-American communities of the Southern United States. Its roots lie in the adoption by African-Americans of European harmony and form, taking on those European elements and combining them into their existing African-based music. Its African musical basis is evident in its use of blue notes, improvisation, polyrhythms, syncopation and the swung note.[1] From its early development until the present day, jazz has also incorporated elements from popular music especially, in its early days, from American popular music.[2]

As the music has developed and spread around the world it has, since its early American beginnings, drawn on many different national, regional and local musical cultures, giving rise to many distinctive styles: New Orleans jazz dating from the early 1910s, big band swing, Kansas City jazz and Gypsy jazz from the 1930s and 1940s, bebop from the mid-1940s on down through Afro-Cuban jazz, West Coast jazz, ska jazz, cool jazz, Indo jazz, avant-garde jazz, soul jazz, modal jazz, chamber jazz, free jazz, Latin jazz in various forms, smooth jazz, jazz fusion and jazz rock, jazz funk, loft jazz, punk jazz, acid jazz, ethno jazz, jazz rap, cyber jazz, M-Base, nu jazz and other ways of playing the music.

Talking of swing, Louis Armstrong, one of the most famous musicians in jazz, said to Bing Crosby on the latter's radio show, "Ah, swing, well, we used to call it syncopation, then they called it ragtime, then blues, then jazz. Now, it's swing. White folks - yo'all sho is a mess!"[3] [4]

In a 1988 interview, trombonist J. J. Johnson said, "Jazz is restless. It won't stay put and it never will".[5]

Definitions

Jazz spans a range of music from ragtime to the present day—a period of over 100 years—and has proved to be very difficult to define. Attempts have been made to define jazz from the perspective of other musical traditions—using the point of view of European music history or African music for example—but critic Joachim Berendt argues that its terms of reference and its definition should be broader.[6] Berendt defines jazz as a "form of art music which originated in the United States through the confrontation of the Negro with European music"[7] and argues that it differs from European music in that jazz has a "special relationship to time defined as 'swing'", involves "a spontaneity and vitality of musical production in which improvisation plays a role" and contains a "sonority and manner of phrasing which mirror the individuality of the performing jazz musician".[6]

A broader definition that encompasses all of the radically different eras of jazz has been proposed by Travis Jackson: he states that it is music that includes qualities such as swing, improvising, group interaction, developing an 'individual voice', and being open to different musical possibilities.[8] An overview of the discussion on definitions is provided by Krin Gabbard, who argues that "jazz is a construct" that, while artificial, still is useful to designate "a number of musics with enough in common to be understood as part of a coherent tradition".[9] In contrast to the efforts of commentators and enthusiasts of certain types of jazz, who have argued for narrower definitions that exclude other types, the musicians themselves are often reluctant to define the music they play. Duke Ellington, one of jazz's most famous figures, summed up this perspective by saying, "It's all music".[10]

Double bassist Reggie Workman, saxophone player Pharaoh Sanders, and drummer Idris Muhammad performing in 1978

Importance of improvisation

While jazz is considered difficult to define, improvisation is consistently regarded as being one of its key elements. The centrality of improvisation in jazz is attributed to its presence in influential earlier forms of music: the early blues, a form of folk music which arose in part from the work songs and field hollers of the African-American workers on plantations. These were commonly structured around a repetitive call-and-response pattern, but early blues was also highly improvisational. Although European classical music has been said to be a composer's medium in which the performer is sometimes granted discretion over interpretation, ornamentation and accompaniment, the performer's primary goal is to play a composition as it was written. In contrast, jazz is often characterized as the product of group creativity, interaction, and collaboration, that places varying degrees of value on the contributions of composer (if there is one) and performers.[11] Summarizing the difference, pianist Earl Hines remarked in a 1975 film that,

> ... when I was playing classical music I wouldn't dare get away from what I was reading. If you've noticed, all of the symphonic musicians, they have played some of those classical tunes for years but they wouldn't vary from one note—and every time they play they have to have the music. So that's why for some classical musicians, it's very difficult for them to try to learn how to play jazz.[12]

In jazz, therefore, the skilled performer will interpret a tune in very individual ways, never playing the same composition exactly the same way twice. Depending upon the performer's mood and personal experience, interactions with other musicians, or even members of the audience, a jazz musician may alter melodies, harmonies or time signature at will. The importance of improvisation has led some critics to suggest that even Duke Ellington's music was not jazz, because it was arranged and orchestrated.[] On the other hand, the solo piano "transformative versions" of Ellington compositions by Earl Hines[13] were described by Ben Ratliff, the *New York Times* jazz critic, as being "as good an example of the jazz process as anything out there".[14]

The approach to improvisation has developed enormously over the history of the music. In early New Orleans and Dixieland jazz, performers took turns playing the melody, while others improvised countermelodies. By the swing era, big bands were coming to rely more on arranged music: arrangements were either written or learned by ear and memorized, while individual soloists would improvise within these arrangements. Later, in bebop the focus shifted back towards small groups and minimal arrangements; the melody (known as the "head") would be stated briefly at the start and end of a piece, but the core of the performance would be the series of improvisations. Later styles such as modal jazz abandoned the strict notion of a chord progression, allowing the individual musicians to improvise even more freely within the context of a given scale or mode. In many forms of jazz a soloist is often supported by a rhythm section that accompanies the soloist by playing chords and rhythms that outline the song structure and complement the soloist.[15] In avant-garde and free jazz idioms, the separation of soloist and band is reduced, and there is license, or even a requirement, for the abandoning of chords, scales, and rhythmic meters.

Debates

Forms of jazz that are commercially oriented or influenced by popular music have been criticized since at least the emergence of bebop. According to Bruce Johnson, there has always been a "tension between jazz as a commercial music and an art form".[8] Traditional jazz enthusiasts have dismissed bebop, free jazz, the 1970s jazz fusion era, and much else as periods of debasement of the music and betrayals of the tradition; the alternative viewpoint is that jazz is able to absorb and transform influences from diverse musical styles,[16] and that, by avoiding the creation of 'norms', other newer, avant-garde forms of jazz will be free to emerge.[8]

Another debate that gained a lot of attention at the birth of jazz was how it would affect the appearance of African-Americans, in particular, who were a part of it. To some African-Americans, jazz has highlighted their contribution to American society and helped bring attention to black history and culture, but for others, the music and term 'jazz' are reminders of "an oppressive and racist society and restrictions on their artistic visions".[17]

Etymology

The origin of the word *jazz* has had wide spread interest—the American Dialect Society named it the Word of the Twentieth Century—which has resulted in considerable research, and its history is well documented. The word began [under various spellings] as West Coast slang around 1912, the meaning of which varied but did not refer to music. The use of the word in a musical context was documented as early as 1915 in the *Chicago Daily Tribune.*[18] Its first documented use in a musical context in New Orleans appears in a November 14, 1916 *Times-Picayune* article about "jas bands."[19]

Race

Jazz has controversial racial connections. Imamu Amiri Baraka argues that there is a distinct "white jazz" music genre expressive of whiteness.[20] The first white jazz musicians appeared in the early 1920s in the Midwestern United States.[21] Bix Beiderbecke was one of the most prominent white jazz musicians.[22]

History

Origins

Blending European and sub-Saharan African music sensibilities

By 1808 the Atlantic slave trade had brought almost half a million Sub-Saharan Africans to the United States. The slaves largely came from West Africa and the greater Congo River basin. They brought strong musical traditions with them.[23] The rhythms had a counter-metric structure, and reflected African speech patterns. African music was largely functional, for work or ritual.[24] The African traditions made use of a single-line melody and call-and-response pattern, but without the European concept of harmony.

Slave gatherings

Lavish festivals featuring African-based dances to drums were organized on Sundays at *Place Congo*, or Congo Square, in New Orleans until 1843.[25] There are historical accounts of other music and dance gatherings elsewhere in the southern United States. Robert Palmer states:

Dance in Congo Square in the late 1700s, artist's conception by E. W. Kemble from a century later.

> An exhaustive analysis of diaries, letters, and travelers' journals from colonial times up to the Civil War, undertaken by Dena J. Epstein and detailed in her book *Sinful Tunes and Spirituals* [1977], yielded a surprising number of references to slave music that was primarily percussive. Usually such music was associated with annual festivals, when the year's crop was harvested and several days were set aside for celebration. As late as 1861, a traveler in North Carolina saw dancers dressed in costumes that included horned headdresses and cow tails and heard music provided by a sheepskin-covered "gumbo box", apparently a frame drum; triangles and jawbones furnished the auxiliary percussion. There are quite a few [accounts] from the southeastern states and Louisiana dating from the period 1820–1850. Some of the earliest [Mississippi] Delta settlers came from the vicinity of New Orleans, where drumming was never actively discouraged for very long and homemade drums were used to accompany public dancing until the outbreak of the Civil War.[26]

In the late 18th-century painting *The Old Plantation*, African-Americans dance to banjo and percussion.

The Black church

Another influence came from black slaves who had learned the harmonic style of hymns of the church, and incorporated it into their own music as spirituals.[27] The origins of the blues are undocumented, though they can be seen as the secular counterpart of the spirituals. However, as Gerhard Kubik points out, whereas the spirituals are homophonic, rural blues and early jazz "was largely based on concepts of heterophony."[28]

Minstrel and salon music

In the early 19th century an increasing number of black musicians learned to play European instruments, particularly the violin, which they used to parody European dance music in their own cakewalk dances. In turn, European-American minstrel show performers in blackface popularized such music internationally, combining syncopation with European harmonic accompaniment. Paul Oliver has drawn attention to similarities in instruments, music and social function to the griots of Africa's western Sudanic belt.[] In the mid-1800s the white New Orleans composer Louis Moreau Gottschalk adapted slave rhythms and melodies from Cuba and other Caribbean islands, into piano salon music. New Orleans was the main nexus between the Afro-Caribbean and African American cultures.

The blackface Virginia Minstrels in 1843, featuring tambourine, fiddle, banjo and bones.

African rhythmic retention

In the opinion of jazz historian Ernest Borneman, what preceded New Orleans jazz before 1890, was "Afro-Latin music" similar to what was played in the Caribbean at the time.[29] A fundamental rhythmic figure heard in Gottschalk's compositions such as "Souvenirs From Havana" (1859), many different slave musics of the Caribbean, as well as the bamboula, and other Afro-Caribbean folk dances performed in New Orleans Congo Square, is the three-stroke pattern known in Cuban music as tresillo. Tresillo is the most basic and by far, the most prevalent duple-pulse rhythmic cell in sub-Saharan African music traditions, and the music of the African Diaspora.[30] [31]

Tresillo.[32] [33] Play

The "Black Codes" outlawed drumming by slaves. Therefore, unlike in Cuba, Haiti, and elsewhere in the Caribbean, African drumming traditions were not preserved in North America. African-based rhythmic patterns were retained in the United States in large part, through "body rhythms" such as stomping, clapping, and patting juba. Robert Palmer states: "*The patting*, an ex-slave reported in 1853, 'is performed by striking the right shoulder with one hand, the left hand with the other—all while keeping time with the feet, and singing.'" African Americans also used everyday household items as percussion instruments. Anthropologist David Evans did extensive fieldwork in the hill country of northern Mississippi, and reports of black families playing polyrhythmic music in their homes on chairs, tin cans, and empty bottles.[34]

There are examples of tresillo, or tresillo-like rhythms in a few surviving nineteenth-century African American folk musics, such as patting juba, and the clapping and foot stomping patterns in ring shout. Palmer describes the foot-generated music:

> Accounts ... leave little doubt that the dancing and stamping constituted a kind of drumming, especially when the worshippers had a wooden church floor to stamp on. "It always rouses my imagination," wrote Lydia Parrish of the Georgia Sea Islands in 1942, "to see the way in which the McIntosh County 'shouters' tap their heels on the resonant floor to imitate the beat of the drum their forebears were not allowed to have."[35] See: The Ringshout and the Birth of African-American Religion [36]

Two decades after drumming was banned in Congo Square, in the post-Civil War period (after 1865), African Americans were able to obtain surplus military bass drums, snare drums and fifes. As a result, an original African American drum and fife music arose, featuring tresillo and related syncopated rhythmic figures.[37] With this music genre, we see the emergence of a drumming tradition that is distinct from its Caribbean counterparts, expressing a sensibility that is uniquely African American. Evans states that among the older black drum and fife musicians of northern Mississippi, making the drums "talk it"—that is, playing rhythm patterns that conform to proverbial phrases or the words of popular fife and drum tunes—"is considered the sign of a good drummer."[38] Palmer observes: "The snare and bass drummers played syncopated cross-rhythms," and speculates—"this tradition must have dated back to the latter half of the nineteenth century, and it could have not have developed in the first place if there hadn't been a reservoir of polyrhythmic sophistication in the culture it nurtured."[38] See: African-American Fife & Drum Music: Mississippi [39].

Tresillo is heard prominently in New Orleans second line music, and in most every other form of popular music to come out of that city from the turn of the twentieth century to present.[40] Jazz historian Gunther Schuller states:

> It is probably safe to say that by and large the simpler African rhythmic patterns survived in jazz ... because they could be adapted more readily to European rhythmic conceptions. Some survived, others were discarded as the Europeanization progressed. It may also account for the fact that patterns such as [tresillo have] ... remained one of the most useful and common syncopated patterns in jazz (1968: 19).[41]

"Spanish tinge"—the Afro-Cuban rhythmic influence

African American music began incorporating Afro-Cuban rhythmic motifs in the nineteenth century, when the habanera (Cuban contradanza) gained international popularity.[42] Habaneras were widely available as sheet music. The habanera was the first written music to be rhythmically based on an African motif (1803).[43] From the perspective of African American music, the *habanera rhythm* (also known as *congo*,[44] *tango-congo*,[45] or *tango*.[46]) can be thought of as a combination of tresillo and the backbeat.[47]

A candombe scene in Montevideo, 1870.

Habanera rhythm written as a combination of tresillo (bottom notes) with the backbeat (top note). Play

Musicians from Havana and New Orleans would take the twice-daily ferry between both cities to perform and not surprisingly, the habanera quickly took root in the musically fertile Crescent City. The habanera was the first of many Cuban music genres which enjoyed periods of popularity in the United States, and reinforced and inspired the use of tresillo-based rhythms in African American music.

John Storm Roberts states that the musical genre habanera, "reached the U.S. twenty years before the first rag was published."[48] The piano piece "Ojos Criollos (Danse Cubaine)" (1860) by New Orleans native Louis Moreau Gottschalk, was influenced by the composer's studies in Cuba. The habanera rhythm is clearly heard in the left hand.[49] With Gottschalk's symphonic work "A Night in the Tropics" (1859), we hear the tresillo variant cinquillo extensively.[50] The figure was also used by Scott Joplin and other ragtime composers.

Cinquillo. Play

For the more than quarter-century in which the cakewalk, ragtime, and proto-jazz were forming and developing, the habanera was a consistent part of African American popular music.[51] Comparing the music of New Orleans with the music of Cuba, Wynton Marsalis observes that tresillo is the New Orleans "clave", a Spanish word meaning 'code,' or 'key'—as in the key to a puzzle, or mystery.[52] Although technically, the pattern is only half a clave, Marsalis makes the important point that the single-celled figure is the guide-pattern of New Orleans music. Jelly Roll Morton called the rhythmic figure the *Spanish tinge*, and considered it an essential ingredient of jazz.[53]

1890s–1910s

Ragtime

Scott Joplin in 1903

The abolition of slavery led to new opportunities for the education of freed African Americans. Although strict segregation limited employment opportunities for most blacks, many were able to find work in entertainment. Black musicians were able to provide "low-class" entertainment in dances, minstrel shows, and in vaudeville, by which many marching bands formed. Black pianists played in bars, clubs, and brothels, as ragtime developed.[54] [55]

Ragtime appeared as sheet music, popularized by African American musicians such as the entertainer Ernest Hogan, whose hit songs appeared in 1895; two years later Vess Ossman recorded a medley of these songs as a banjo solo "Rag Time Medley".[56] [57] Also in 1897, the white composer William H. Krell published his "Mississippi Rag" as the first written piano instrumental ragtime piece, and Tom Turpin published his Harlem Rag, that was the first rag published by an African-American.

The classically trained pianist Scott Joplin and the acknowledged "king of ragtime" produced his "Original Rags" in the following year, then in 1899 had an international hit with "Maple Leaf Rag". "Maple Leaf Rag" is a multi-strain ragtime march with athletic bass lines and offbeat melodies. Each of the four parts features a recurring theme and a striding bass line with copious seventh chords. The piece may be considered the 'archetypal rag' due to its influence on the genre; its structure was the basis for many other famous rags, including "Sensation" by Joseph Lamb. It is more carefully constructed than almost all the previous rags, and the syncopations in the right hand, especially in the transition between the first and second strain, were novel at the time.[58]

Excerpt from "Maple Leaf Rag" by Scott Joplin (1899). Seventh chord resolution. Play. Note that the seventh resolves down by half step.

African-based rhythmic patterns, such as tresillo, and its variants—the habanera rhythm and cinquillo, are heard in the ragtime compositions of Scott Joplin, Tom Turpin, and others. Joplin's "Solace" (1909) is generally considered to be within the habanera genre (although it's labeled a "Mexican serenade").[44] [59] The following excerpt from "Solace" is based on two different variants of the habanera rhythm.

Excerpt from "Solace" by Scott Joplin (1909). Variations on the habanera rhythm.

With "Solace" both hands are playing in a syncopated fashion, completely abandoning any sense of a march rhythm. Ned Sublette postulates that the tresillo/habanera rhythm "found its way into ragtime and the cakewalk,"[60] while Roberts suggests that "the habanera influence may have been part of what freed black music from ragtime's European bass."[61]

Joplin wrote numerous popular rags, including "The Entertainer", combining right hand tresillo-based syncopation, banjo figurations and sometimes call-and-response. The ragtime idiom was eventually taken up by classical composers including Claude Debussy and Igor Stravinsky.

Blues

African genesis

Blues is the name given to both a musical form and a music genre[62] that originated in African-American communities of primarily the "Deep South" of the United States at the end of the 19th century from spirituals, work songs, field hollers, shouts and chants, and rhymed simple narrative ballads.[] The African use of pentatonic scales contributed to the development of blue notes in blues and jazz.[63] As Kubik explains:

> Many of the rural blues of the Deep South are *stylistically* an extension and merger of basically two broad accompanied song-style traditions in the west central Sudanic belt:

- A strongly Arabic/Islamic song style, as found for example among the Hausa. It is characterized by melisma, wavy intonation, pitch instabilities within a pentatonic framework, and a declamatory voice.
- An ancient west central Sudanic stratum of pentatonic song composition, often associated with simple work rhythms in a regular meter, but with notable off-beat accents (1999: 94).[64]

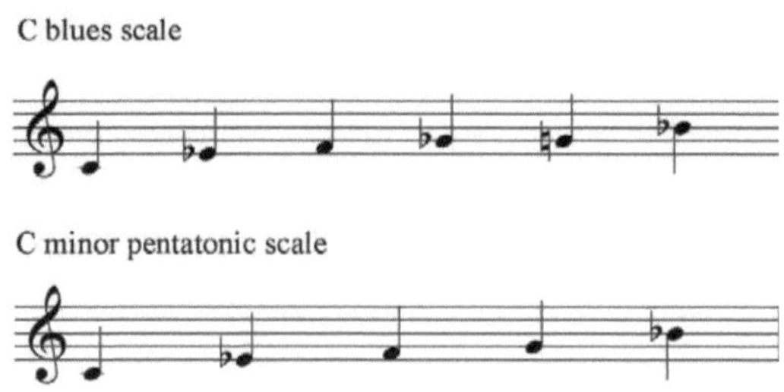

Play blues scale or pentatonic scale

Within the context of Western harmony

In 1892 St. Louis, Missouri, W.C. Handy, an out of work African American cornet player, with experience in minstrel shows and brass bands, encountered his first blues (or proto-blues) song. It had numerous one-line verses "and they would sing it all night."[65] In 1912, Handy published what he heard that night as "St. Louis Blues." In 1903, while traveling through the Mississippi Delta, Handy experienced a form of blues with more pronounced African traits. The Delta blues style intrigued him. The singer improvised freely, and the melodic range was limited, sounding like a field holler. The guitar accompaniment was not strummed, but was instead, like a small drum that responded in syncopated accents. The guitar was another "voice."[66] Handy's "St. Louis Blues" and "Memphis Blues" (1912) are jazz standards.[] While many identify Handy's "Memphis Blues" as the first published blues, Gunther Schuller argues that it is not really a blues, but "more like a cakewalk."[67]

The blues form, ubiquitous in jazz, is characterized by specific chord progressions, of which the twelve-bar blues chord progression is the most common. The blue notes that, for expressive purposes are sung or played flattened or gradually bent (minor 3rd to major 3rd) in relation to the pitch of the major scale, are also an important part of the sound. The blues were the key that opened up an entirely new approach to Western harmony, ultimately leading to a high level of harmonic complexity in jazz.

New Orleans

The Bolden Band around 1905.

The music of New Orleans had a profound effect on the creation of early jazz. Many early jazz performers played in venues throughout the city; the brothels and bars of the red-light district around Basin Street, called "Storyville"[68] was only one of numerous neighborhoods relevant to the early days of New Orleans jazz. In addition to dance bands, numerous marching bands played at lavish funerals arranged by the African American and European American community. The instruments used in marching bands and dance bands became the basic instruments of jazz: brass and reeds tuned in the European 12-tone scale and drums. Small bands mixing self-taught and well educated African American musicians, many of whom came from the funeral-procession tradition of New Orleans, played a seminal role in the development and dissemination of early jazz, traveling throughout Black communities in the Deep South and, from around 1914 on, Afro-Creole and African American musicians playing in vaudeville shows took jazz to western and northern US cities.[]

Syncopation

The cornetist Buddy Bolden led a band often mentioned as one of the prime movers of the style later to be called "jazz". He played in New Orleans around 1895–1906. Bolden's band is credited with creating the *big four*, the first syncopated bass drum pattern to deviate from the standard on-the-beat march.[69] As the example below shows, the second half of the big four pattern is the habanera rhythm.

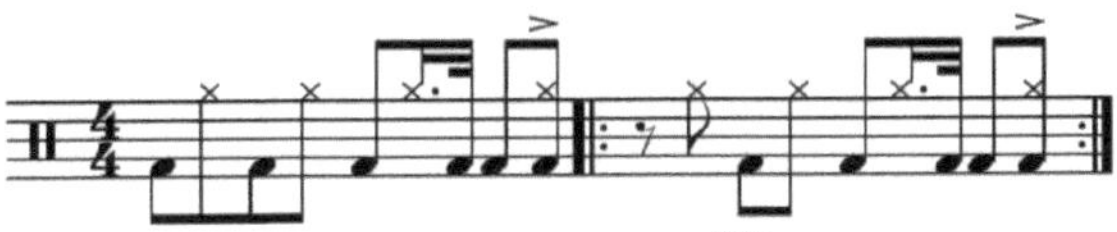
Buddy Bolden's "big four" pattern.[70] Play

No recordings remain of Bolden. Several tunes from the Bolden band repertory, including "Buddy Bolden Blues", have been recorded by many other musicians. Bolden became mentally ill and spent his later decades in a mental institution.

Morton published "Jelly Roll Blues" in 1915, the first jazz work in print.

Afro-Creole pianist Jelly Roll Morton began his career in Storyville. From 1904, he toured with vaudeville shows around southern cities, also playing in Chicago and New York. His "Jelly Roll Blues", which he composed around 1905, was published in 1915 as the first jazz arrangement in print, introducing more musicians to the New Orleans style.[71]

Morton would perform habaneras, such as "La Paloma." He considered the tresillo/habanera (which he called the Spanish tinge) to be an essential ingredient of jazz.[72] The habanera rhythm and tresillo can be heard in his left hand on songs like "The Crave" (1910, recorded 1938). In Morton's own words:

> "Now in one of my earliest tunes, "New Orleans Blues," you can notice the Spanish tinge. In fact, if you can't manage to put tinges of Spanish in your tunes, you will never be able to get the right seasoning, I call it, for jazz"—Morton (1938: Library of Congress Recording).[53]

Excerpt from Jelly Roll Morton's "New Orleans Blues" (c. 1902). The left hand plays the tresillo rhythm. The right hand plays variations on cinquillo. Play

Some early jazz musicians referred to their music as *ragtime*. Jelly Roll Morton was a crucial innovator in the evolution from ragtime to jazz piano. Morton could perform pieces in either style.[73] Morton's solos were still close to ragtime, and were not merely improvisations over chord changes, as with later jazz. His use of the blues was of equal importance however.

Swing

Morton loosened ragtime's rhythmic feeling, decreasing its embellishments, and employing a swing feeling.[74] Swing is the most important, and enduring African-based rhythmic technique used in jazz. An oft quoted definition of swing by Louis Armstrong is: "if you don't feel it, you'll never know it."[75] *The New Harvard Dictionary of Music* states that swing is: "An intangible rhythmic momentum in jazz . . . Swing defies analysis; claims to its presence may inspire arguments." However, the dictionary does provide the useful description of triple subdivisions of the beat contrasted with duple subdivisions.[76] Swing superimposes six subdivisions of the beat over a basic pulse structure or four subdivisions. This aspect of swing is far more prevalent in African American music than in Afro-Caribbean music. One aspect of swing, which is heard in more rhythmically complex Diaspora musics, places strokes in-between the triple and duple-pulse "grids."[77]

Bottom: even duple subdivisions of the beat. Top: swung correlative—contrasting of duple and triple subdivisions of the beat. Play straight drum pattern or Play swung pattern

New Orleans brass bands are a lasting influence contributing horn players to the world of professional jazz with the distinct sound of the city while helping black children escape poverty.[78] The leader of the Camelia Brass Band, D'Jalma Ganier, taught Louis Armstrong to play trumpet. Armstrong popularized the New Orleans style of trumpet playing, and then expanded it. Like Jelly Roll Morton, Louis Armstrong is also credited with the abandonment of ragtime's stiffness, in favor of swung notes. Armstrong, perhaps more than any other musician, codified the rhythmic technique of swing in jazz, and broadened the jazz solo vocabulary.[79]

The Original Dixieland Jass Band made the music's first recordings early in 1917, and their "Livery Stable Blues" became the earliest released jazz record.[80] [81] [82] [83] [84] [85] [86] That year numerous other bands made recordings featuring "jazz" in the title or band name, mostly ragtime or novelty records rather than jazz. In February 1918 James Reese Europe's "Hellfighters" infantry band took ragtime to Europe during World War I,[87] then on return recorded Dixieland standards including "Darktown Strutters' Ball".[]

Other regions

WC Handy age 19, 1892

W.C. Handy, who was from Alabama, states that when he was travelling with his band in 1896, he was playing "novelty music ... similar to jazz, but we didn't call it jazz."[88] Later, in 1903, Handy became intrigued with the folk blues of the Deep South. He and his band members were formally trained African American musicians, who did not grow up with the blues. Yet, Handy was able to adopt the blues to a larger band instrument format, and arrange them in a popular music form. Handy wrote about his adopting of the blues:

> "The primitive southern Negro, as he sang, was sure to bear down on the third and seventh tone of the scale, slurring between major and minor. Whether in the cotton field of the Delta or on the Levee up St. Louis way, it was always the same. Till then, however, I had never heard this slur used by a more sophisticated Negro, or by any white man. I tried to convey this effect ... by introducing flat thirds and sevenths (now called blue notes) into my song, although its prevailing key was major ..., and I carried this device into my melody as well ... This was a distinct departure, but as it turned out, it touched the spot."[89]

The 1912 publication of his "Memphis Blues" sheet music introduced the 12-bar blues to the world. Handy's autobiography is titled *Father of the Blues*.[90] Like Morton, Handy's music career began in the pre-jazz era, and contributed to the codification of jazz, through the publication of some of the first jazz sheet music. In September 1917 Handy's Orchestra of Memphis recorded a cover version of "Livery Stable Blues."[91]

Also like Morton, Handy performed habaneras. Handy noted a reaction to the habanera rhythm included in Will H. Tyler's "Maori"—"I observed that there was a sudden, proud and graceful reaction to the rhythm ... White dancers, as I had observed them, took the number in stride. I began to suspect that there was something Negroid in that beat."

After noting a similar reaction to the same rhythm in "La Paloma", Handy included this rhythm in his "St. Louis Blues" (1914).

Excerpt from "St. Louis Blues" by W.C. Handy (1914). The left hand plays the habanera rhythm.

In addition, Handy used the habanera rhythm in the instrumental copy of "Memphis Blues", the chorus of "Beale Street Blues", and other compositions."[92]

In the northeastern United States, a "hot" style of playing ragtime had developed, notably James Reese Europe's symphonic Clef Club orchestra in New York which played a benefit concert at Carnegie Hall in 1912.[] [93] The Baltimore rag style of Eubie Blake influenced James P. Johnson's development of "Stride" piano playing, in which the right hand plays the melody, while the left hand provides the rhythm and bassline.[94]

In Ohio and elsewhere in the midwest, ragtime was the major influence until about 1919. Around 1912, when the four-string banjo and saxophone came in, the musicians began to improvise the melody line, but the harmony and rhythm remained unchanged. A contemporary account states that blues could only be heard in jazz, in the gut-bucket cabarets, which were generally looked down upon by the Black middle-class.[95]

1920s and 1930s

The Jazz Age

Trumpeter, bandleader and singer Louis Armstrong was a much-imitated innovator of early jazz.

Prohibition in the United States (from 1920 to 1933) banned the sale of alcoholic drinks, resulting in illicit speakeasies becoming lively venues of the "Jazz Age", an era when popular music included current dance songs, novelty songs, and show tunes. Jazz started to get a reputation as being immoral and many members of the older generations saw it as threatening the old values in culture and promoting the new decadent values of the Roaring 20s. Professor Henry van Dyke of Princeton University wrote "... it is not music at all. It's merely an irritation of the nerves of hearing, a sensual teasing of the strings of physical passion."[]

Even the media began to denigrate jazz. The New York Times took stories and altered headlines to pick at jazz. For instance, villagers used pots and pans in Siberia to scare off bears, and the newspaper stated that it was jazz that scared the bears away. Another story claims that Jazz caused the death of a celebrated conductor. The actual cause of death was a fatal heart attack (natural cause).[]

From 1919 Kid Ory's Original Creole Jazz Band of musicians from New Orleans played in San Francisco and Los Angeles where in 1922 they became the first black jazz band of New Orleans origin to make recordings.[96] [] However, the main center developing the new "Hot Jazz" was Chicago, where King Oliver joined Bill Johnson. That year also saw the first recording by Bessie Smith, the most famous of the 1920s blues singers.[97] Bix Beiderbecke formed The Wolverines in 1924.

The King & Carter Jazzing Orchestra photographed in Houston, Texas, January 1921.

Also in 1924 Louis Armstrong joined the Fletcher Henderson dance band as featured soloist for a year. The original New Orleans style was polyphonic, with theme variation, and simultaneous collective improvisation. Armstrong was a master of his hometown style, but by the time he joined Henderson's band, he was already a trailblazer in a new phase of jazz, with its emphasis on arrangements and soloists. Armstrong's solos went well beyond the theme-improvisation concept, and extemporized on chords, rather than melodies. According to Schuller, by comparison, the solos by Armstrong's bandmates (including a young Coleman Hawkins), sounded "stiff, stodgy," with "jerky rhythms and a grey undistinguished tone quality."[98] The following example shows a short excerpt of the straight melody of "Mandy, Make Up Your Mind" by Irving Berlin (top), compared with Louis Armstrong's solo improvisations (below) (recorded 1924).[99] The example approximates Armstrong's solo, as it doesn't convey his use of swing.

Top: excerpt from the straight melody of "Mandy, Make Up Your Mind" by Irving Berlin. Bottom: corresponding solo excerpt by Louis Armstrong (1924).

Armstrong's solos were a significant factor in making jazz a true twentieth-century language. After leaving Henderson's group, Armstrong formed his virtuosic Hot Five band, where he popularized scat singing.[100]

Jelly Roll Morton recorded with the New Orleans Rhythm Kings in an early mixed-race collaboration, then in 1926 formed his Red Hot Peppers. There was a larger market for jazzy dance music played by white orchestras, such as Jean Goldkette's orchestra and Paul Whiteman's orchestra. In 1924 Whiteman commissioned Gershwin's *Rhapsody in Blue*, which was premiered by Whiteman's Orchestra. Other influential large ensembles included Fletcher Henderson's band, Duke Ellington's band (which opened an influential residency at the Cotton Club in 1927) in New York, and Earl Hines's Band in Chicago (who opened in The Grand Terrace Cafe there in 1928). All significantly influenced the development of big band-style swing jazz.[101] By 1930, the New Orleans-style ensemble was a relic, and jazz belonged to the world.[102]

Swing

Benny Goodman (1943)

The 1930s belonged to popular swing big bands, in which some virtuoso soloists became as famous as the band leaders. Key figures in developing the "big" jazz band included bandleaders and arrangers Count Basie, Cab Calloway, Jimmy and Tommy Dorsey, Duke Ellington, Benny Goodman, Fletcher Henderson, Earl Hines, Glenn Miller and Artie Shaw.

Swing was also dance music. It was broadcast on the radio 'live' nightly across America for many years especially by Hines and his Grand Terrace Cafe Orchestra broadcasting[103] coast-to-coast from Chicago, well placed for 'live' time-zones. Although it was a collective sound, swing also offered individual musicians a chance to 'solo' and improvise melodic, thematic solos which could at times be very complex and 'important' music.

Over time, social strictures regarding racial segregation began to relax in America: white bandleaders began to recruit black musicians and black bandleaders white ones.

In the mid-1930s, Benny Goodman hired pianist Teddy Wilson, vibraphonist Lionel Hampton and guitarist Charlie Christian to join small groups. An early 1940s style known as "jumping the blues" or jump blues used small combos, uptempo music, and blues chord progressions. Jump blues drew on boogie-woogie from the 1930s. Kansas City Jazz in the 1930s as exemplified by tenor saxophonist Lester Young marked the transition from big bands to the bebop influence of the 1940s.

Beginnings of European jazz

Outside of the United States the beginnings of a distinct European style of jazz emerged in France with the Quintette du Hot Club de France, which began in 1934. Belgian guitar virtuoso Django Reinhardt popularized gypsy jazz, a mix of 1930s American swing, French dance hall "musette" and Eastern European folk with a languid, seductive feel. The main instruments are steel stringed guitar, violin, and double bass. Solos pass from one player to another as the guitar and bass play the role of the rhythm section. Some music researchers hold that it was Philadelphia's Eddie Lang and Joe Venuti who pioneered the guitar-violin partnership typical of the genre,[104] which was brought to France after they had been heard live or on Okeh Records in the late 1920s.[105]

1940s and 1950s

"American music"—the genius of Ellington

Duke Ellington at the Hurricane Club (1943)

By the 1940s, Duke Ellington's music transcended the bounds of swing, bridging jazz and art music in a natural synthesis. Ellington called his music "American Music" rather than jazz, and liked to describe those who impressed him as "beyond category."[106] These included many of the musicians who were members of his orchestra, some of whom are considered among the best in jazz in their own right, but it was Ellington who melded them into one of the most well-known jazz orchestral units in the history of jazz. He often composed specifically for the style and skills of these individuals, such as "Jeep's Blues" for Johnny Hodges, "Concerto for Cootie" for Cootie Williams, which later became "Do Nothing Till You Hear from Me" with Bob Russell's lyrics, and "The Mooche" for Tricky Sam Nanton and Bubber Miley. He also recorded songs written by his bandsmen, such as Juan Tizol's "Caravan" and "Perdido" which brought the "Spanish Tinge" to big-band jazz. Several members of the orchestra remained there for several decades. The band reached a creative peak in the early 1940s, when Ellington and a small hand-picked group of his composers and arrangers wrote for an orchestra of distinctive voices who displayed tremendous creativity.[107]

Bebop

Thelonious Monk at Minton's Playhouse, 1947, New York City.

In the early 1940s bebop-style performers began to shift jazz from danceable popular music towards a more challenging "musician's music." The most influential bebop musicians included saxophonist Charlie Parker, pianists Bud Powell and Thelonious Monk, trumpeters Dizzy Gillespie and Clifford Brown, and drummer Max Roach. Composer Gunther Schuller wrote:

> ... In 1943 I heard the great Earl Hines band which had Bird in it and all those other great musicians. They were playing all the flatted fifth chords and all the modern harmonies and substitutions and Dizzy Gillespie runs in the trumpet section work. Two years later I read that that was 'bop' and the beginning of modern jazz ... but the band never made recordings.[108]

Divorcing itself from dance music, bebop established itself more as an art form, thus lessening its potential popular and commercial appeal. Dizzy Gillespie wrote:

> ... People talk about the Hines band being 'the incubator of bop' and the leading exponents of that music ended up in the Hines band. But people also have the erroneous impression that the music was new. It was not. The music evolved from what went before. It was the same basic music. The difference was in how you got from here to here to here ... naturally each age has got its own shit.[109]

Earl Hines 1947

Rhythm

Since bebop was meant to be listened to, not danced to, it could use faster tempos. Drumming shifted to a more elusive and explosive style, in which the ride cymbal was used to keep time while the snare and bass drum were used for accents. This led to a highly syncopated, linear rhythmic complexity.[110]

Harmony

Charlie Parker, Tommy Potter, Miles Davis, Max Roach (Gottlieb 06941)

Bebop musicians employed several harmonic devices not typical of previous jazz, engaging in a more abstracted form of chord-based improvisation. Bebop scales are traditional scales, with an added chromatic passing note.[111] Bebop also uses "passing" chords, substitute chords, and altered chords. New forms of chromaticism and dissonance were introduced into jazz; the dissonant tritone (or "flatted fifth") interval became the "most important interval of bebop"[112] Chord progressions for bebop tunes were often taken directly from popular swing-era songs and reused with a new and more complex melody, forming new compositions. This practice was already well-established in earlier jazz, but came to be central to the bebop style. Bebop made use of several relatively common chord progressions, such as blues (at base, I-IV-V, but infused with II-V motion) and 'rhythm changes' (I-VI-II-V), the chords to the 1930s pop standard "I Got Rhythm." Late bop also moved towards extended forms that represented a departure from pop and show tunes. The harmonic development in bebop, is often traced back to a transcendent moment experienced by Charlie Parker while performing "Cherokee" at Clark Monroe's Uptown House, New York, in early 1942.

> I'd been getting bored with the stereotyped changes that were being used, ... and I kept thinking there's bound to be something else. I could hear it sometimes. I couldn't play it.... I was working over 'Cherokee,' and, as I did, I found that by using the higher intervals of a chord as a melody line and backing them with appropriately related changes, I could play the thing I'd been hearing. It came alive—Parker.[113]

Gerhard Kubik postulates that the harmonic development in bebop sprang from the blues, and other African-related tonal sensibilities, rather than twentieth century Western art music, as some have suggested. Kubik states: "Auditory inclinations were the African legacy in [Parker's] life, reconfirmed by the experience of the blues tonal system, a sound world at odds with the Western diatonic chord categories. Bebop musicians eliminated Western-style functional harmony in their music while retaining the strong central tonality of the blues as a basis for drawing upon various African matrices."[114] Samuel Floyd states that blues were both the bedrock and propelling force of bebop, bringing about three main developments:

- A new harmonic conception, using extended chord structures that led to unprecedented harmonic and melodic variety.
- A developed and even more highly syncopated, linear rhythmic complexity and a melodic angularity in which the blue note of the fifth degree was established as an important melodic-harmonic device.
- The reestablishment of the blues as the music's primary organizing and functional principle.[110]

> While for an outside observer, the harmonic innovations in bebop would appear to be inspired by experiences in Western "serious" music, from Claude Debussy to Arnold Schoenberg, such a scheme cannot be sustained by the evidence from a cognitive approach. Claude Debussy did have some influence on jazz, for example, on Bix Beiderbecke's piano playing. And it is also true that Duke Ellington adopted and reinterpreted some harmonic devices in European contemporary music. West Coast jazz would run into such debts as would several forms of cool jazz, but bebop has hardly any such debts in the sense of direct borrowings. On the contrary, ideologically, bebop was a strong statement of rejection of any kind of eclecticism, propelled by a desire to activate something deeply buried in self. Bebop then revived tonal-harmonic ideas transmitted through the blues and reconstructed and expanded others in a basically non-Western harmonic approach. The ultimate significance of all this is that the experiments in jazz during the 1940s brought back to African-American music several structural principles and techniques rooted in African traditions—Kubik (2005).[115]

These divergences from the jazz mainstream of the time initially met with a divided, sometimes hostile, response among fans and fellow musicians, especially established swing players, who bristled at the new harmonic sounds. To hostile critics, bebop seemed to be filled with "racing, nervous phrases".[116] Despite the initial friction, by the 1950s bebop had become an accepted part of the jazz vocabulary.

Afro-Cuban jazz (cu-bop)

Machito and Mario Bauza

Machito (maracas) and his sister Graciella Grillo (claves)

The general consensus among musicians and musicologists is that the first original jazz piece to be overtly based *in-clave* was "Tanga" (1943), composed by Cuban-born Mario Bauza and recorded by Machito and his Afro-Cubans in New York City. "Tanga" began as a spontaneous descarga (Cuban jam session) with jazz solos superimposed on top.[117]

This was the birth of Afro-Cuban jazz. The use of clave brought the African *timeline*, or *key pattern*, into jazz. Music organized around key patterns convey a two-celled (binary) structure, which is a complex level of African cross-rhythm.[118] Within the context of jazz however, harmony is the primary referent, not rhythm. The harmonic progression can begin on either side of clave, and the harmonic "one" is always understood to be "one". If the progression begins on the "three-side" of clave, it is said to be in *3-2 clave*. If the progression begins on the "two-side", its in *2-3 clave*.[119]

Clave: Spanish for 'code,' or key,' as in the key to a puzzle. The antecedent half (three-side) consists of tresillo. The consequent half consists of two strokes (the two-side). Play

Bobby Sanabria mentions several innovations of Machito's Afro-Cubans; they were the first band to: wed big band jazz arranging techniques within an original composition, with jazz oriented soloists utilizing an authentic Afro-Cuban based rhythm section in a successful manner; explore modal harmony (a concept explored much later by Miles Davis and Gil Evans) from a jazz arranging perspective; and to *overtly* explore the concept of clave conterpoint from an arranging standpoint (the ability to weave seamlessly from one side of the clave to the other without breaking its rhythmic integrity within the structure of a musical arrangement). They were also the first band in the United States to publicly utilize the term *Afro-Cuban* as the band's moniker, thus identifying itself and acknowledging the West African roots of the musical form they were playing. It forced New York City's Latino and African American communities to deal with their common West African musical roots in a direct way, whether they wanted to acknowledge it publicly or not.[120]

Dizzy Gillespie and Chano Pozo

Dizzy Gillespie, 1955

Mario Bauzá introduced bebop innovator Dizzy Gillespie to the Cuban conga drummer, dancer, composer, and choreographer Chano Pozo. The brief collaboration of Gillespie and Pozo produced some of the most enduring Afro-Cuban jazz standards. "Manteca" (1947), co-written by Gillespie and Pozo, is the first jazz standard to be rhythmically based on clave. According to Gillespie, Pozo composed the layered, contrapuntal guajeos (Afro-Cuban ostinatos) of the A section and the introduction, while Gillespie wrote the bridge. Gillespie recounted: "If I'd let it go like [Chano] wanted it, it would have been strickly Afro-Cuban all the way. There wouldn't have been a bridge. I thought I was writing an eight-bar bridge, but after eight bars I hadn't resolved back to B-flat, so I had to keep going and ended up writing a sixteen-bar bridge."[121]

It was the bridge that gave "Manteca" a typical jazz harmonic structure, setting the piece apart from Bauza's modal "Tanga" of a few years earlier. Jazz arrangements with a "Latin" A section and a swung B section, with all choruses swung during solos, became common practice with many "Latin tunes" of the jazz standard repertoire. This approach can be heard on pre-1980 recordings of "Manteca", "A Night in Tunisia", "Tin Tin Deo", and "On Green Dolphin Street."

Gillespie's collaboration with Pozo brought specific African-based rhythms into bebop. While pushing the boundaries of harmonic improvisation, *cu-bop* as it was called, also drew more directly from African rhythmic structures. The rhythm of the melody of the A section is identical to a common mambo bell pattern.

Top: opening measures of "Manteca" melody. Bottom: common mambo bell pattern (2–3 clave).

African cross-rhythm

Mongo Santamaria (1969)

Cuban percussionist Mongo Santamaria first recorded his composition "Afro Blue" in 1959.[122] "Afro Blue" was the first jazz standard built upon a typical African three-against-two (3:2) cross-rhythm, or hemiola.[123] The song begins with the bass repeatedly playing 6 cross-beats per each measure of 12/8, or 6 cross-beats per 4 main beats—6:4 (two cells of 3:2). The following example shows the original ostinato "Afro Blue" bass line. The slashed noteheads indicate the main beats (not bass notes), where you would normally tap your foot to "keep time."

"Afro Blue" bass line, with main beats indicated by slashed noteheads.

When John Coltrane covered "Afro Blue" in 1963, he inverted the metric hierarchy, interpreting the tune as a 3/4 jazz waltz with duple cross-beats superimposed (2:3). Originally a Bb pentatonic blues, Coltrane expanded the harmonic structure of "Afro Blue."

Perhaps the most respected Afro-cuban jazz combo of the late 1950s was vibraphonist Cal Tjader's band. Tjader had Mongo Santamaria, Armando Peraza, and Willie Bobo on his early recording dates.

Dixieland revival

In the late 1940s there was a revival of "Dixieland" music, harkening back to the original contrapuntal New Orleans style. This was driven in large part by record company reissues of early jazz classics by the Oliver, Morton, and Armstrong bands of the 1930s. There were two types of musicians involved in the revival. One group consisted of players who had begun their careers playing in the traditional style and were returning to it or continuing what they had been playing all along. This included Bob Crosby's Bobcats, Max Kaminsky, Eddie Condon, and Wild Bill Davison.[124] Most of this group were originally Midwesterners, although there were a small number of New Orleans musicians involved. The second group of revivalists consisted of younger musicians, such as those in the Lu Watters band. By the late 1940s, Louis Armstrong's Allstars band became a leading ensemble. Through the 1950s and 1960s, Dixieland was one of the most commercially popular jazz styles in the US, Europe, and Japan, although critics paid little attention to it.[124]

Cool jazz

By the end of the 1940s, the nervous energy and tension of bebop was replaced with a tendency towards calm and smoothness, with the sounds of cool jazz, which favoured long, linear melodic lines. It emerged in New York City, as a result of the mixture of the styles of predominantly white jazz musicians and black bebop musicians, and it dominated jazz in the first half of the 1950s. The starting point was a series of singles on Capitol Records in 1949 and 1950 of a nonet led by trumpeter Miles Davis, collected and released first on a ten-inch and later a twelve-inch as the *Birth of the Cool.* Cool jazz recordings by Chet Baker, Dave Brubeck, Bill Evans, Gil Evans, Stan Getz and the Modern Jazz Quartet usually have a "lighter" sound which avoided the aggressive tempos and harmonic abstraction of bebop.

Cool jazz later became strongly identified with the West Coast jazz scene, but also had a particular resonance in Europe, especially Scandinavia, with emergence of such major figures as baritone saxophonist Lars Gullin and pianist Bengt Hallberg. The theoretical underpinnings of cool jazz were set out by the blind Chicago pianist Lennie Tristano, and its influence stretches into such later developments as Bossa nova, modal jazz, and even free jazz. List of Cool jazz and West Coast jazz musicians for further detail.

Hard bop

Hard bop is an extension of bebop (or "bop") music that incorporates influences from rhythm and blues, gospel music, and blues, especially in the saxophone and piano playing. Hard bop was developed in the mid-1950s, partly in response to the vogue for cool jazz in the early 1950s. The hard bop style coalesced in 1953 and 1954, paralleling the rise of rhythm and blues. Miles Davis' performance of "Walkin'", at the very first Newport Jazz Festival in 1954, announced the style to the jazz world. The quintet Art Blakey and the Jazz Messengers, fronted by Blakey and featuring pianist Horace Silver and trumpeter Clifford Brown, were leaders in the hard bop movement along with

Davis.

Modal jazz

Modal jazz is a development beginning in the later 1950s which takes the mode, or musical scale, as the basis of musical structure and improvisation. Previously, the goal of the soloist was to play a solo that fit into a given chord progression. However, with modal jazz, the soloist creates a melody using one or a small number of modes. The emphasis in this approach shifts from harmony to melody.[125] Pianist Mark Levine states: "Historically, this caused a seismic shift among jazz musicians, away from thinking vertically (the chord), and towards a more horizontal approach (the scale)."[126]

The modal theory stems from a work by George Russell. Miles Davis introduced the concept to the greater jazz world with *Kind of Blue* (1959), an exploration of the possibilities of modal jazz and the best selling jazz album of all time. In contrast to Davis's earlier work with the hard bop style of jazz and its complex chord progression and improvisation,[] the entire album was composed as a series of *modal sketches*, in which each performer was given a set of scales that defined the parameters of their improvisation and style.[] Davis recalled: "I didn't write out the music for *Kind of Blue*, but brought in sketches for what everybody was supposed to play because I wanted a lot of spontaneity."[127] The track "So What" has only two chords: D-7 and E♭-7.[128] Listen: "So What" by Miles Davis (1959). [129]

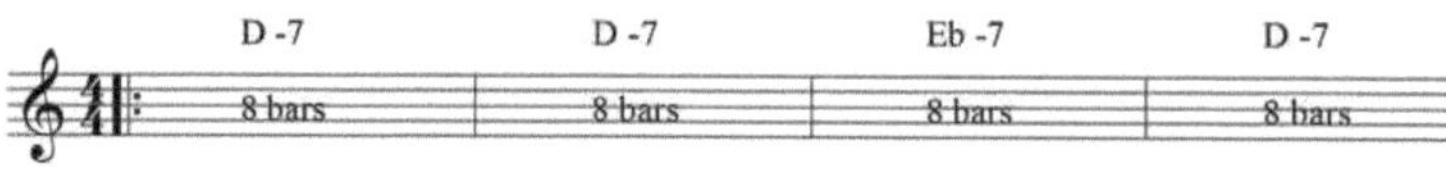

Chord changes for "So What" by Miles Davis (1959).

Other innovators in this style include Jackie McLean,[130] John Coltrane and Bill Evans, also present on *Kind of Blue*, as well as later musicians such as Herbie Hancock. Coltrane's modal "Impressions" is based on the same changes as Miles Davis' "So What."[131] Watch: "Impressions" (1961) by John Coltrane, with Eric Dolphy, McCoy Tyner, Reggie Workman, and Elvin Jones. [132]

By the 1950s, Afro-Cuban jazz had been using modes for at least a decade. This was because a lot of Afro-Cuban jazz borrowed from Cuban popular dance forms, which are structured around multiple ostinatos with only a few chords. A case in point is Mario Bauza's "Tanga" (1943), the first Afro-Cuban jazz piece. Machito's Afro-Cubans recorded modal tunes in the 1940s, featuring jazz soloists such as Howard McGhee, Brew Moore, Charlie Parker, and Flip Phillips. Listen: "Tanga" performed by Machito's Afro-Cubans. NYC c. 1940s. [133] There is no evidence however, that Miles Davis or other mainstream jazz musicians were influenced by the use of modes in Afro-Cuban jazz, or other branches of Latin jazz. By the 1950s Latin jazz was generally considered a novelty by the mainstream, and the genre had a limited influence.

Free jazz

Free jazz and the related form of avant-garde jazz broke through into an open space of "free tonality" in which meter, beat, and formal symmetry all disappeared, and a range of World music from India, Africa, and Arabia were melded into an intense, even religiously ecstatic or orgiastic style of playing.[134] While loosely inspired by bebop, free jazz tunes gave players much more latitude; the loose harmony and tempo was deemed controversial when this approach was first developed. The bassist Charles Mingus is also frequently associated with the avant-garde in jazz, although his compositions draw from myriad styles and genres.

A shot from a 2006 performance by Peter Brötzmann, a key figure in European free jazz

The first major stirrings came in the 1950s, with the early work of Ornette Coleman and Cecil Taylor. In the 1960s, performers included Archie Shepp, Sun Ra, Albert Ayler, Pharaoh Sanders, John Coltrane, and others. In developing his late style, Coltrane was especially influenced by the dissonance of Ayler's trio with bassist Gary Peacock and drummer Sunny Murray, a rhythm section honed with Cecil Taylor as leader. Coltrane championed many younger free jazz musicians, (notably Archie Shepp), and under his influence Impulse! became a leading free jazz record label.

A series of recordings with the Classic Quartet in the first half of 1965 show John Coltrane's playing becoming increasingly abstract, with greater incorporation of devices like multiphonics, utilization of overtones, and playing in the altissimo register, as well as a mutated return to Coltrane's sheets of sound. In the studio, he all but abandoned his soprano to concentrate on the tenor saxophone. In addition, the quartet responded to the leader by playing with increasing freedom. The group's evolution can be traced through the recordings *The John Coltrane Quartet Plays*, *Living Space*, *Transition* (both June 1965), *New Thing at Newport* (July 1965), *Sun Ship* (August 1965), and *First Meditations* (September 1965).

In June 1965, Coltrane and ten other musicians recorded *Ascension*, a 40-minute long piece that included adventurous solos by the young avant-garde musicians (as well as Coltrane), and was controversial primarily for the collective improvisation sections that separated the solos. After recording with the quartet over the next few months, Coltrane invited Pharoah Sanders to join the band in September 1965. While Coltrane used over-blowing frequently as an emotional exclamation-point, Sanders would opt to overblow his entire solo, resulting in a constant screaming and screeching in the altissimo range of the instrument.

Free jazz quickly found a foothold in Europe—in part because musicians such as Ayler, Taylor, Steve Lacy and Eric Dolphy spent extended periods there. A distinctive European contemporary jazz (often incorporating elements of free jazz but not limited to it) flourished also because of the emergence of musicians (such as John Surman, Zbigniew Namyslowski, Albert Mangelsdorff, Kenny Wheeler and Mike Westbrook) anxious to develop new approaches reflecting their national and regional musical cultures and contexts. Keith Jarrett has been prominent in defending free jazz from criticism by traditionalists in the 1990s and 2000s.

1960s and 1970s

Latin jazz

Latin jazz is jazz with Latin American rhythms. Although musicians continually expand its parameters, the term Latin jazz is generally understood to have a more specific meaning than simply jazz from Latin America. A more precise term might be Afro-Latin jazz, as the jazz sub-genre typically employs rhythms that either have a direct analog in Africa, or exhibit an African rhythmic influence beyond what is ordinarily heard in other jazz. The two main categories of Latin jazz are Afro-Cuban jazz and Brazilian jazz.

In the 1960s and 1970s, many jazz musicians had only a minimum understanding of Cuban and Brazilian music. Jazz compositions using Cuban or Brazilian elements were often referred to as "Latin tunes", with no distinction between a Cuban son montuno and a Brazilian bossa nova. Even as late as 2000, in Mark Gridley's *Jazz Styles: History and Analysis*, a bossa nova bass line is referred to as a "Latin bass figure."[135] It was not uncommon during the 1960s and 1970s to hear a conga playing a Cuban tumbao, while the drumset and bass played a Brazilian bossa nova pattern. Many jazz standards such as "Manteca", "On Green Dolphin Street", and "Song for My Father", have a "Latin" A section, and a swung B section. Typically, the band would only play an even-eighth "Latin" feel in the A section of the head and swing throughout all of the solos. Latin jazz specialists like Cal Tjader tended to be the exception. For example, on a 1959 live Tjader recording of "A Night in Tunisia", pianist Vince Guaraldi soloed through the entire form over an authentic mambo.[136]

Afro-Cuban jazz

Afro-Cuban jazz often uses Afro-Cuban instruments such as congas, timbales, güiro, and claves, combined with piano, double bass, etc. Afro-Cuban jazz began with Machito's Afro-Cubans in the early 1940s, but took off and entered the mainstream in the late 1940s when bebop musicians such as Dizzy Gillespie and Billy Taylor began experimenting with Cuban rhythms. Mongo Santamaria and Cal Tjader further refined the genre in the late 1950s. Although a great deal of Cuban-based Latin jazz is modal, Latin jazz is not always modal. It can be as harmonically expansive as post-bop jazz. For example, Tito Puente recorded an arrangement of "Giant Steps" done to an Afro-Cuban guaguancó. A Latin jazz piece may momentarily contract harmonically, as in the case of a percussion solo over a one or two-chord piano guajeo.

Guajeos

Guajeos are the typical Afro-Cuban ostinato melodies, which originated in the genre known as son. Guajeos provide a rhythmic/melodic framework that may be varied within certain parameters, while still maintaining a repetitive, and thus "dancable", structure. Most guajeos are rhythmically based on clave, the organizing principle of a great deal of Afro-Cuban, and sub-Saharan African music. Guajeos, or guajeo fragments are commonly used motifs in Latin jazz compositions.

In addition to common jazz concepts, soloists in Latin jazz draw from the improvisational vocabulary of the Afro-Cuban *descarga* (jazz-inspired instrumental jams), and popular dance forms such as salsa. Guajeos are one of the most important elements of this vocabulary, providing a means of tension/resolution, and a sense of forward momentum, within a relatively simple harmonic structure. The use of multiple, contrapuntal guajeos in Latin jazz facilitates simultaneous collective improvisation, based on theme variation. In a way, this polyphonic texture is reminiscent of the original New Orleans style of jazz.

Afro-Cuban jazz renaissance

Afro-Cuban jazz has been for most of its history a matter of superimposing jazz phrasing over Cuban rhythms. However, by the end of the 1970s, a new generation of New York City musicians emerged who were fluent in both salsa dance music and jazz. The time had come for a new level of integration of jazz and Cuban rhythms. This era of creativity and vitality is best represented by the Gonzalez brothers Jerry (congas and trumpet) and Andy (bass).[137] During 1974-1976 they were members of one of Eddie Palmieri's most experimental salsa groups. Salsa was the medium, but Palmieri was stretching the form in new ways. He incorporated the use of parallel fourths, with McCoy Tyner-type vamps. The innovations of Palmieri, the Gonzalez brothers and others, led to an Afro-Cuban jazz renaissance in New York City, and eventually, worldwide, during the 1980s. Today, jazz musicians are expected to have a general knowledge of clave, and the different genres within Cuban and Brazilian music.

Afro-Brazilan jazz

Naná Vasconcelos playing the Afro-Brazilian Berimbau

Brazilian jazz such as bossa nova is derived from samba, with influences from jazz and other 20th century classical and popular music styles. Bossa is generally moderately paced, with melodies sung in Portuguese or English. The style was pioneered by Brazilians João Gilberto and Antônio Carlos Jobim. The related term jazz-samba describes an adaptation of street samba into jazz. Bossa nova was made popular by Elizete Cardoso's recording of *Chega de Saudade* on the *Canção do Amor Demais* LP. The initial releases by Gilberto and the 1959 film *Black Orpheus* achieved significant popularity in Latin America, and this spread to North America via visiting American jazz musicians. The resulting recordings by Charlie Byrd and Stan Getz cemented bossa nova's popularity and led to a worldwide boom with 1963's *Getz/Gilberto*, numerous recordings by famous jazz performers such as Ella Fitzgerald (*Ella Abraça Jobim*) and Frank Sinatra (*Francis Albert Sinatra & Antônio Carlos Jobim*) and the entrenchment of the bossa nova style as a lasting influence in world music for several decades and even up to the present.

Brazilian percussionists such as Airto Moreira and Naná Vasconcelos also influenced jazz internationally by introducing Afro-Brazilian folkloric instruments and rhythms into a wide variety of jazz styles and attracting a greater audience to them.[138] [139] [140]

Post-bop

Post-bop jazz is a form of small-combo jazz derived from earlier bop styles. The genre's origins lie in seminal work by John Coltrane, Miles Davis, Bill Evans, Charles Mingus, Wayne Shorter and Herbie Hancock. Generally, the term post-bop is taken to mean jazz from the mid-sixties onward that assimilates influence from hard bop, modal jazz, the avant-garde, and free jazz, without necessarily being immediately identifiable as any of the above.

Much post-bop was recorded on Blue Note Records. Key albums include *Speak No Evil* by Wayne Shorter; *The Real McCoy* by McCoy Tyner; *Maiden Voyage* by Herbie Hancock; *Miles Smiles* by Miles Davis; and *Search for the New Land* by Lee Morgan (an artist not typically associated with the post-bop genre). Most post-bop artists worked in other genres as well, with a particularly strong overlap with later hard bop.

Soul jazz

Soul jazz was a development of hard bop which incorporated strong influences from blues, gospel and rhythm and blues in music for small groups, often the organ trio, which partnered a Hammond organ player with a drummer and a tenor saxophonist. Unlike hard bop, soul jazz generally emphasized repetitive grooves and melodic hooks, and improvisations were often less complex than in other jazz styles. Horace Silver had a large influence on the soul jazz style, with songs that used funky and often gospel-based piano vamps. It often had a steadier "funk" style groove, different from the swing rhythms typical of much hard bop. Important soul jazz organists included Jimmy McGriff and Jimmy Smith and Johnny Hammond Smith, and influential tenor saxophone players included Eddie "Lockjaw" Davis and Stanley Turrentine.

African inspired

Themes

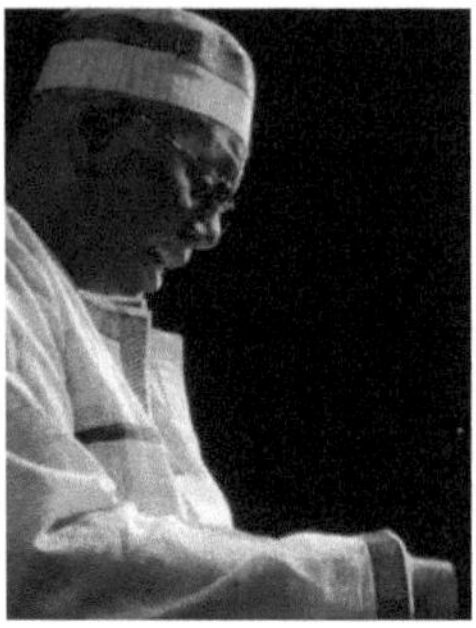

Randy Weston

There was a resurgence of interest in jazz and other forms of African American cultural expression during the Black Arts Movement and Black nationalist period of the 1960s and 1970s. African themes became popular. There were many new jazz compositions with African-related titles: "Black Nile" (Wayne Shorter), "Blue Nile" (Alice Coltrane), "Obirin African" (Art Blakey), "Zambia" (Lee Morgan), "Appointment in Ghana" (Jackie McLean), "Marabi" (Cannonball Adderley), "Yoruba" (Hubert Laws), and many more. Pianist Randy Weston's music incorporated African elements, for example, the large-scale suite "Uhuru Africa" (with the participation of poet Langston Hughes) and "Highlife: Music From the New African Nations." Both Weston and saxophonist Stanley Turrentine covered the Nigerian Bobby Benson's piece "Niger Mambo", which features Afro-Caribbean and jazz elements within a West African Highlife style. Some musicians such as Pharaoh Sanders, Hubert Laws and Wayne Shorter began using African instruments such as kalimbas, bells, beaded gourds and other instruments not traditional to jazz.

Rhythm

During this period, there was an increased use of the typical African 12/8 cross-rhythmic structure in jazz. Herbie Hancock's "Succotash" on *Inventions and Dimensions* (1963) is an open-ended modal, 12/8 jazz-descarga (jam), improvised on the spot, with no written music. Accompanied by Paul Chambers on bass, and the Latin percussionists Willie Bobo and Osvaldo Martinez "Chihuahua", Hancock's pattern of attack-points, rather than the pattern of pitches, is the primary focus of his improvisations. Martinez plays a traditional Afro-Cuban chekeré part, while Bobo plays an Abakuá bell pattern on a snare drum with brushes.

Abakuá bell pattern played on a snare with brushes by Willie Bobo on Herbie Hancock's "Succotash" (1963).

The first jazz standard composed by a non-Latino to use an overt African 12/8 cross-rhythm was Wayne Shorter's "Footprints" (1967).[141] On the version recorded on *Miles Smiles* by Miles Davis, the bass switches to 4/4 at 2:20. The 4/4 figure is known as tresillo in Latin music and is the duple-pulse correlative of the cross-beats in triple-pulse. "Footprints" is not, however, a Latin jazz tune; Cuban music is not serving as the conduit to African rhythmic structures. Those structures are accessed directly by Ron Carter (bass) and Tony Williams (drums), via the rhythmic sensibilities of swing. Throughout the piece, the four beats, whether sounded or not, are maintained as the temporal referent. In the example below the main beats are indicated by slashed noteheads. They are shown here for reference, and do not indicate bass notes.

Ron Carter's two main bass lines for "Footprints" by Wayner Shorter (1967). The main beats are indicated by slashed noteheads.

Pentatonic scales

The use of pentatonic scales was another African-associated trend. The use of pentatonic scales in Africa probably goes back thousands of years.[142] McCoy Tyner perfected the use of the pentatonic scale in his solos.[143] Tyner also used parallel fifths and fourths, which are common harmonies in West Africa.[144]

The minor pentatonic scale is often used in blues improvisation. Like a blues scale, a minor pentatonic scale can be played over all of the chords in a blues. The following pentatonic lick was played over blues changes by Joe Henderson on Horace Silver's "African Queen" (1965).[145]

C minor pentatonic phrase played by Joe Henderson on "African Queen" by Horace Silver (1965).

Jazz pianist, theorist, and educator Mark Levine refers the scale generated by beginning on the fifth step of a pentatonic scale, as the *V pentatonic scale*.[146]

C pentatonic scale beginning on the I (C pentatonic), IV (F pentatonic), and V (G pentatonic) steps of the scale.

Levine points out that the V pentatonic scale works for all three chords of the standard II-V-I jazz progression.[147] This is a very common progression, used in pieces such as Miles Davis' "Tune Up." The following example shows the V pentatonic scale over a II-V-I progression.[148]

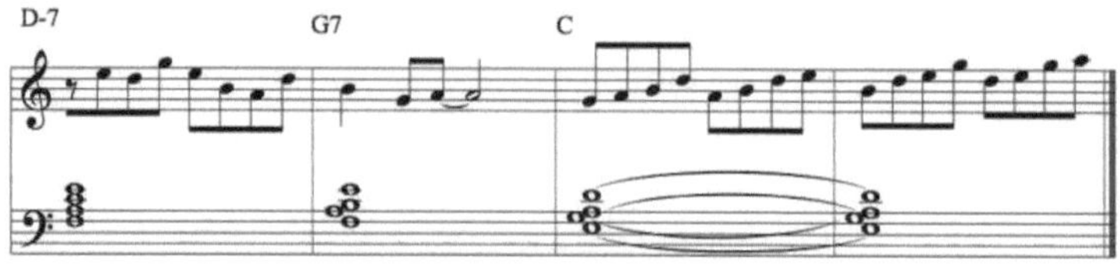

V pentatonic scale over II-V-I chord progression.

Accordingly, John Coltrane's "Giant Steps" (1960), with its 26 chords per 16 bars, can be played using only three pentatonic scales. Coltrane studied Nicolas Slonimsky's *Thesaurus of Scales and Melodic Patterns*, which contains material that is virtually identical to portions of "Giant Steps".[149] The harmonic complexity of "Giant Steps" is on the level of the most advanced twentieth-century art music. Superimposing the pentatonic scale over "Giant Steps" is not merely a matter of harmonic simplification, but also a sort of "Africanizing" of the piece, which provides an alternate approach for soloing. Mark Levine observes that when mixed in with more conventional "playing the changes", pentatonic scales provide "structure and a feeling of increased space."[150]

Jazz fusion

In the late 1960s and early 1970s the hybrid form of jazz-rock fusion was developed by combining jazz improvisation with rock rhythms, electric instruments and the highly amplified stage sound of rock musicians such as Jimi Hendrix. Jazz fusion music often uses mixed meters, odd time signatures, syncopation, complex chords and harmonies. All Music Guide states that "until around 1967, the worlds of jazz and rock were nearly completely separate. [However, ...] as rock became more creative and its musicianship improved, and as some in the jazz world became bored with hard bop and did not want to play strictly avant-garde music, the two different idioms began to trade ideas and occasionally combine forces."[151]

Fusion trumpeter Miles Davis in 1989

Miles Davis' new directions

In 1969 Davis fully embraced the electric instrument approach to jazz with *In a Silent Way*, which can be considered his first fusion album. Composed of two side-long suites edited heavily by producer Teo Macero, this quiet, static album would be equally influential upon the development of ambient music. As Davis recalls: "The music I was really listening to in 1968 was James Brown, the great guitar player Jimi Hendrix, and a new group who had just come out with a hit record, "Dance to the Music," Sly and the Family Stone... I wanted to make it more like rock. When we recorded In a Silent Way I just threw out all the chord sheets and told everyone to play off of that."[152] *In a Silent Way* featured contributions from musicians who would all go on to spread the fusion evangel with their own groups in the 1970s: Shorter, Hancock, Corea, pianist Josef Zawinul, John McLaughlin, Holland, and Williams. Williams quit Davis to form the group The Tony Williams Lifetime with McLaughlin and organist Larry Young. Their debut record of that year *Emergency!* is also cited as one of the early acclaimed fusion albums.

Psychedelic-jazz

Bitches Brew

Davis's *Bitches Brew* (1970) was his most successful of this era. Although inspired by rock and funk, Davis's fusion creations were original, and brought about a type of new avant-garde, electronic, psychedelic-jazz, as far from pop music as any other Davis work.

Herbie Hancock

Davis alumnus, pianist Herbie Hancock, released four albums of the short-lived (1970–1973) psychedelic-jazz sub-genre: *Mwandishi* (1972), *Crossings* (1973), and *Sextant* (1973). The rhythmic background was a mix of rock, funk, and African-type textures.

Musicians who worked with Davis formed the four most influential fusion groups: Weather Report and Mahavishnu Orchestra emerged in 1971 and were soon followed by Return to Forever and The Headhunters.

Weather Report

Weather Report's debut album was in the electronic, psychedelic-jazz vein. The self-titled *Weather Report* (1971) caused a sensation in the jazz world on its arrival, thanks to the pedigree of the group's members (including percussionist Airto Moreira), and their unorthodox approach to their music. The album featured a softer sound than would be the case in later years (predominantly using acoustic bass, with Shorter exclusively playing soprano saxophone, and with no synthesizers involved) but is still considered a classic of early fusion. It built on the avant-garde experiments which Zawinul and Shorter had pioneered with Miles Davis on *Bitches Brew* (including an avoidance of head-and-chorus composition in favour of continuous rhythm and movement) but taking the music further. To emphasise the group's rejection of standard methodology, the album opened with the inscrutable avant-garde atmospheric piece "Milky Way" (created by Shorter's extremely muted saxophone inducing vibrations in Zawinul's piano strings while the latter pedalled the instrument). Down Beat described the album as "music beyond category" and awarded it Album of the Year in the magazine's polls that year. Weather Report's subsequent releases were creative funk-jazz works.[153]

Jazz-rock

Although jazz purists protested the blend of jazz and rock, some of jazz's significant innovators crossed over from the contemporary hard bop scene into fusion. In addition to using the electric instruments of rock, such as the electric guitar, electric bass, electric piano and synthesizer keyboards, fusion also used the powerful amplification, "fuzz" pedals, wah-wah pedals, and other effects used by 1970s-era rock bands. Notable performers of jazz fusion included Miles Davis, keyboardists Joe Zawinul, Chick Corea, Herbie Hancock, vibraphonist Gary Burton, drummer Tony Williams, violinist Jean-Luc Ponty, guitarists Larry Coryell, Al Di Meola, John McLaughlin and Frank Zappa, saxophonist Wayne Shorter and bassists Jaco Pastorius and Stanley Clarke. Jazz fusion was also popular in Japan where the band Casiopea released over thirty fusion albums.

In the twenty-first century, almost all jazz has influences from other nations and styles of music, making jazz fusion as much a common practice as style.

Jazz-funk

Developed by the mid-1970s, jazz-funk is characterized by a strong back beat (groove), electrified sounds,[154] and often, the presence of electronic analog synthesizers. The integration of funk, soul and R&B music into jazz resulted in the creation of a genre whose spectrum is wide and ranges from strong jazz improvisation to soul, funk or disco with jazz arrangements, jazz riffs and jazz solos, and sometimes soul vocals.[]

Early examples are Herbie Hancock's Headhunters band and the Miles Davis album *On the Corner*. The latter, from 1972, began Miles Davis's foray into jazz-funk and was, Davis claimed, an attempt at reconnecting with the young black audience which had largely forsaken jazz for rock and funk. While there is a discernible rock and funk influence in the timbres of the instruments employed, other tonal and rhythmic textures, such as the Indian tambora and tablas, and Cuban congas and bongos, create a multi-layered soundscape. From a musical standpoint, the album was a culmination of sorts of the *musique concrète* approach that Davis and producer Teo Macero had begun to explore in the late 1960s.

At the jazz end of the spectrum, jazz-funk characteristics include a departure from ternary rhythm (near-triplet), i.e., the "swing", to the more danceable and unfamiliar binary rhythm, known as the "groove". Jazz-funk also draws influences from traditional African music, Afro-Cuban rhythms and Jamaican reggae, most notably Kingston band leader Sonny Bradshaw. A second characteristic of jazz-funk is the use of electric instruments. The ARP Odyssey, ARP String Ensemble and Hohner D6 Clavinet also became popular at the time. A third feature is the shift of proportions between composition and improvisation: arrangements, melody and overall writing were heavily emphasized.

Irakere and the emergence of the Cuban school

The first Cuban band of this new wave was Irakere (1973). "Chékere-son" (1976) introduced a style of "Cubanized" bebop-flavored horn lines, that departed from the more angular guajeo-based lines typical of Cuban popular music, and Latin jazz up until that time. It was based on Charlie Parker's bebop composition "Billie's Bounce". Almost all of "Billie's Bounce" can be found in "Chékere-son", but it is jumbled together in a way that fuses clave and bebop horn lines.[155]

The horn line style introduced in "Chékere-son" is heard today in Afro-Cuban jazz, and the contemporary popular dance genre known as timba. Another important Irakere contribution is their use of batá and other Afro-Cuban folkloric drums. Ironically, several of the founding members did not always appreciate Irakere's fusion of jazz and Afro-Cuban elements. They saw the Cuban folk elements as a type of nationalistic "fig leaf", cover for their true love—jazz. In spite of the ambivalence by some members towards Irakere's Afro-Cuban folkloric/jazz fusion, their experiments forever changed Cuban jazz: their innovations are heard in the high level of harmonic and rhythmic complexity in Cuban jazz, and in the jazzy and complex contemporary form of popular dance music known as timba.

Other trends

Musicians began improvising jazz tunes on unusual instruments, such as the jazz harp (Alice Coltrane), electrically amplified and wah-wah pedaled jazz violin (Jean-Luc Ponty), and even bagpipes (Rufus Harley). Jazz continued to expand and change, influenced by other types of music, such as world music, avant garde classical music, and rock and pop music. Guitarist John McLaughlin's Mahavishnu Orchestra played a mix of rock and jazz infused with East Indian influences. The ECM record label began in Germany in the 1970s with artists including Keith Jarrett, Paul Bley, the Pat Metheny Group, Jan Garbarek, Ralph Towner, Kenny Wheeler, John Taylor, John Surman and Eberhard Weber, establishing a new chamber music aesthetic, featuring mainly acoustic instruments, and sometimes incorporating elements of world music and folk music.

1980s

In 1987, the US House of Representatives and Senate passed a bill proposed by Democratic Representative John Conyers, Jr. to define jazz as a unique form of American music stating, among other things, "... that jazz is hereby designated as a rare and valuable national American treasure to which we should devote our attention, support and resources to make certain it is preserved, understood and promulgated." It passed in the House of Representatives on September 23, 1987 and in the Senate on November 4, 1987.[156]

Resurgence of Traditionalism

Wynton Marsalis

While the 1970s had been dominated by the fusion and free jazz genres, the early 1980s saw a re-emergence of a more conventional kind of acoustic or straight-ahead jazz. Perhaps the most prominent manifestation of this resurgence was the emergence of trumpeter Wynton Marsalis, who strove to create music within what he believed was the tradition, rejecting both fusion and free jazz and creating extensions of the small and large forms initially pioneered by such artists as Louis Armstrong and Duke Ellington as well as the hard bop of the 1950s. Several musicians who had been prominent in the fusion genre during the 1970s began to record acoustic jazz once more, including Chick Corea and Herbie Hancock. Even the early-80s music of Miles Davis, although still recognisably fusion, adopted a far more conventional approach than his abstract work of the 1970s. A similar reaction took place against free jazz: according to Ted Giola,

> the very leaders of the avant garde started to signal a retreat from the core principles of Free Jazz. Anthony Braxton began recording standards over familiar chord changes. Cecil Taylor played duets in concert with Mary Lou Williams, and let her set out structured harmonies and familiar jazz vocabulary under his blistering keyboard attack. And the next generation of progressive players would be even more accommodating, moving inside and outside the changes without thinking twice. Musicians such as David Murray or Don Pullen may have felt the call of free-form jazz, but they never forgot all the other ways one could play African-American music for fun and profit.[157]

Smooth jazz

David Sanborn, 2008

In the early 1980s, a commercial form of jazz fusion called pop fusion or "smooth jazz" became successful and garnered significant radio airplay. Smooth jazz saxophonists include Grover Washington, Jr., Kenny G, Kirk Whalum, Boney James and David Sanborn. Smooth jazz received frequent airplay with more straight-ahead jazz in "quiet storm" time slots at radio stations in urban markets across the U.S., helping to establish or bolster the careers of vocalists including Al Jarreau, Anita Baker, Chaka Khan and Sade. In this same time period Chaka Khan released *Echoes of an Era*, which featured Joe Henderson, Freddie Hubbard, Chick Corea, Stanley Clarke and Lenny White. She also released the song "And the Melody Still Lingers On (Night in Tunisia)" with Dizzy Gillespie reviving the solo break from "Night in Tunisia".

In general, smooth jazz is downtempo (the most widely played tracks are in the 90–105 BPM range), layering a lead, melody-playing instrument (saxophones—especially soprano and tenor—are the most popular, with legato electric guitar playing a close second). In his *Newsweek* article "The Problem With Jazz Criticism"[158] Stanley Crouch considers Miles Davis' playing of fusion as a turning point that led to smooth jazz. In Aaron J. West's introduction to his analysis of smooth jazz, "Caught Between Jazz and Pop" he states,

> I challenge the prevalent marginalization and malignment of smooth jazz in the standard jazz narrative. Furthermore, I question the assumption that smooth jazz is an unfortunate and unwelcomed evolutionary outcome of the jazz-fusion era. Instead, I argue that smooth jazz is a long-lived musical style that merits multi-disciplinary analyses of its origins, critical dialogues, performance practice, and reception.[159]

Acid jazz, nu jazz and jazz rap

Gang Starr in Hamburg, Germany, 1999

Acid jazz developed in the UK over the 1980s and 1990s, influenced by jazz-funk and electronic dance music. Jazz-funk musicians such as Roy Ayers and Donald Byrd are often credited as forerunners of acid jazz.[160] While acid jazz often contains various types of electronic composition (sometimes including sampling or live DJ cutting and scratching), it is just as likely to be played live by musicians, who often showcase jazz interpretation as part of their performance. Nu jazz is influenced by jazz harmony and melodies; there are usually no improvisational aspects. It ranges from combining live instrumentation with beats of jazz house, exemplified by St Germain, Jazzanova and Fila Brazillia, to more band-based improvised jazz with electronic elements such as that of The Cinematic Orchestra, Kobol, and the Norwegian "future jazz" style pioneered by Bugge Wesseltoft, Jaga Jazzist, Nils Petter Molvær, and others. Nu jazz can be very experimental in nature and can vary widely in sound and concept.

Jazz rap developed in the late 1980s and early 1990s, and incorporates jazz influence into hip hop. In 1988, Gang Starr released the debut single "Words I Manifest", sampling Dizzy Gillespie's 1962 "Night in Tunisia", and Stetsasonic released "Talkin' All That Jazz", sampling Lonnie Liston Smith. Gang Starr's debut LP, *No More Mr. Nice Guy* (Wild Pitch, 1989), and their track "Jazz Thing" (CBS, 1990) for the soundtrack of *Mo' Better Blues*, sampling Charlie Parker and Ramsey Lewis. Gang Starr also collaborated with Branford Marsalis and Terence Blanchard.Groups making up the collective known as the Native Tongues Posse tended towards jazzy releases; these include the Jungle Brothers' debut *Straight Out the Jungle* (Warlock, 1988) and A Tribe Called Quest's *People's Instinctive Travels and the Paths of Rhythm* (Jive, 1990) and *The Low End Theory* (Jive, 1991).

The Low End Theory has become one of hip hop's most acclaimed albums, and earned praise too from jazz bassist Ron Carter, who played double bass on one track. Rap duo Pete Rock & CL Smooth incorporated jazz influences on their 1992 debut Mecca and the Soul Brother. Beginning in 1993, rapper Guru's Jazzmatazz series used jazz musicians during the studio recordings. Though jazz rap had achieved little mainstream success, jazz legend Miles Davis' final album (released posthumously in 1992), *Doo-Bop*, was based around hip hop beats and collaborations with producer Easy Mo Bee. Davis' ex-bandmate Herbie Hancock returned to hip hop influences in the mid-nineties, releasing the album *Dis Is Da Drum* in 1994.

Punk jazz and jazzcore

John Zorn performing in 2006

The relaxation of orthodoxy concurrent with post-punk in London and New York City led to a new appreciation for jazz. In London, the Pop Group began to mix free jazz, along with dub reggae, into their brand of punk rock.[161] In NYC, No Wave took direct inspiration from both free jazz and punk. Examples of this style include Lydia Lunch's *Queen of Siam*,[162] the work of James Chance and the Contortions, who mixed Soul with free jazz and punk,[162] Gray, and the Lounge Lizards,[162] who were the first group to call themselves "punk jazz."

John Zorn began to make note of the emphasis on speed and dissonance that was becoming prevalent in punk rock and incorporated this into free jazz. This began in 1986 with the album *Spy vs. Spy*, a collection of Ornette Coleman tunes done in the contemporary thrashcore style.[163] The same year, Sonny Sharrock, Peter Brötzmann, Bill Laswell, and Ronald Shannon Jackson recorded the first album under the name Last Exit, a similarly aggressive blend of thrash and free jazz.[164] These developments are the origins of *jazzcore*, the fusion of free jazz with hardcore punk.

In the 1990s, punk jazz and jazzcore began to reflect the increasing awareness of elements of extreme metal (particularly thrash metal and death metal) in hardcore punk. A new style of "metallic jazzcore" was developed by Iceburn, from Salt Lake City, and Candiria, from New York City, though anticipated by Naked City and Pain Killer. This tendency also takes inspiration from jazz inflections in technical death metal, such as the work of Cynic and Atheist.

M-Base

Steve Coleman in Paris, July 2004

The M-Base movement was started in the 1980s by a loose collective of young African-American musicians (Steve Coleman, Graham Haynes, Cassandra Wilson, Geri Allen, Greg Osby etc.) who emerged in New York with a new sound and specific ideas about creative expression. With a strong foothold as well as in the tradition represented by Charlie Parker and John Coltrane as in contemporary African-American groove music and with a high degree of musical skills,[165] the saxophonists Steve Coleman, Greg Osby, and Gary Thomas developed unique and complex, nevertheless grooving[166] musical languages. In the 1990s most participants of the M-Base movement turned to more conventional music but Steve Coleman, the most active participant, continued developing his music in accordance with the M-Base concept.[167]

In a long research process he developed a philosophical and spiritual concept connecting with certain cultural efforts that express fundamental aspects of nature and human existence in a holistic way. Steve Coleman found these efforts all over the world and they reach far back into ancient times.[167] Thus, he gave his music a specific meaning which is similar to the intentions of religious music, of European composers like J.S. Bach and Ludwig van Beethoven, as well as of musicians in the tradition represented by John Coltrane.[168] Coleman's audience decreased but his music and concepts influenced many musicians[169] —both in terms of music-technique[170] and of the music's meaning.[171] Hence, "M-Base" changed from a movement of a loose collective of young musicians to a kind of informal Steve Coleman "school"[172] with a much advanced but already originally implied concept.[173]

1990s–2010s

Jazz since the 1990s has been characterised by a pluralism in which no one style dominates but rather a wide range of active styles and genres are popular. Often individual performers will record and play music in a variety of different styles, sometimes in the same performance. Musicians emerging since the 1990s and usually performing in more-or-less straight-ahead settings include US pianists Brad Mehldau, Jason Moran and Vijay Iyer, guitarist Kurt Rosenwinkel, vibraphonist Stefon Harris, trumpeters Roy Hargrove and Terence Blanchard, saxophonists Chris Potter and Joshua Redman and bassist Christian McBride.

Although jazz-rock fusion reached the height of its popularity in the 1970s, the use of electronic instruments and rock-derived musical elements in jazz continued in the 1990s and 2000s. Musicians using this approach have included Pat Metheny, John Abercrombie, John Scofield and Swedish group e.s.t. among others. Pianist Brad Mehldau and power trio The Bad Plus have explored contemporary rock music within the context of the traditional jazz acoustic piano trio, for example recording instrumental jazz versions of songs by rock musicians. The Bad Plus have also incorporated elements of free jazz into their music. A number of new vocalists have achieved popularity with a mix of traditional jazz and pop/rock forms, such as Diana Krall, Norah Jones, Cassandra Wilson, Kurt Elling and Jamie Cullum.

While some musicians such as saxophonists Greg Osby and Charles Gayle have maintained a firm avant-garde or free jazz stance, others such as James Carter have incorporated free jazz elements into a more traditional framework. Established musicians in a variety of styles continued to perform, such as Wayne Shorter, Herbie Hancock, Chick Corea, Sonny Rollins, Wynton Marsalis, Keith Jarrett, David Murray and McCoy Tyner.

See also

- Cape jazz
- Glossary of jazz and popular music
- Jazz poetry
- Timeline of jazz education
- Victorian Jazz Archive

Lists:

- List of jazz bassists
- List of jazz clubs
- List of jazz drummers
- List of jazz festivals
- List of jazz guitarists
- List of jazz institutions and organizations
- List of jazz pianists
- List of jazz standards
- List of jazz violinists
- List of jazz vocalists

Notes

[1] Alyn Shipton, *A New History of Jazz*, 2nd ed., Continuum, 2007, pp. 4–5

[2] Bill Kirchner, *The Oxford Companion to Jazz*, Oxford University Press, 2005, Chapter Two.

[4] Father of the Blues by William Christopher Handy. 1941 MacMillan page 292

[5] J J Johnson continued, "[Jazz] is forever seeking and reaching out and exploring": DownBeat: *The Great Jazz Interviews – A 75th Anniversary Anthology*: p. 250

[6] Joachim E. Berendt. *The Jazz Book: From Ragtime to Fusion and Beyond*. Translated by H. and B. Bredigkeit with Dan Morgenstern. 1981. Lawrence Hill Books. Page 371

[7] Berendt, Joachim Ernst (1964) *The New Jazz Book: a History and Guide*, p. 278. P. Owen (http://books.google.es/books?ei=8Rz-UdDVDefb7AbhvYCYBA&id=pjgYAQAAIAAJ&dq=Joachim+Berendt+which+originated+in+the+United+States+through+the+confrontation+of+blacks&q=confrontation#search_anchor) At Google Books. Retrieved 4 August 2013.

[8] In Review of *The Cambridge Companion to Jazz* (http://www.fzmw.de/2003/2003_12.htm) by Peter Elsdon, FZMw (Frankfurt Journal of Musicology) No. 6, 2003

[11] Giddins 1998 70.

[12] *Earl "Fatha" Hines*: 1 hr docu (https://vimeo.com/58414566) – see External Links, [Charlie Nairn] below.

[13] On *Earl Hines Plays Duke Ellington* recorded in the 1970s

[14] Ratliff 2002, 19.

[15] Jazz Drum Lessons (http://www.drumbook.org/drum-lessons/jazz-drums/jazz-drumming/) – Drumbook.org

[16] In "Jazz Inc." (http://www.metrotimes.com/music/features/19/12/jazz.html) by Andrew Gilbert, *Metro Times*, December 23, 1998

[18] Archived at Observatoire Musical Français, Paris-Sorbonne University.

[24] African musician/scholar Kofi Agawu disputes this conventional view: "The idea that African music is functional in contrast to a contemplative European music is a myth ... this particular construct arose in connection with earlier ethnographies written for a Western audience and aiming to convey what might be different about African music. The point that African music can be legitimately listened to, that it never relinguishes a contemplative dimension—not even in theory—apparrently still needs to be made in view of long-standing views linking music to dance." Agawu, Kofi (2003: 104–105). *Representing African Music: Postcolonial Notes, Queries, Positions*. New York: Routledge.

[25] "The primary instrument for a cultural music expression was a long narrow African drum. It came in various sized from three to eight feet long and had previously been banned in the South by whites. Other instruments used were the triangle, a jawbone, and early ancestors to the banjo. Many types of dances were performed in Congo Square, including the 'flat-footed-shuffle' and the 'Bamboula.'" *African American Registry*. http://www.aaregistry.org/historic_events/view/congo-square-soul-new-orleans

[26] Palmer, Robert (1981: 37). *Deep Blues*. New York: Penguin.

[28] Kubik, Gerhard (1999: 112).

[29] Borneman, Ernest (1969: 104). Jazz and the Creole Tradition." *Jazz Research* I: 99–112.

[30] Sublette, Ned (2008: 124, 287). *The World that made New Orleans: from Spanish silver to Congo Square*. Chicago: Lawrence Hill Books. ISBN 1-55652-958-9

[31] Peñalosa, David (2010: 38–46). *The Clave Matrix; Afro-Cuban Rhythm: Its Principles and African Origins*. Redway, CA: Bembe Inc. ISBN 1-886502-80-3.

[32] Garrett, Charles Hiroshi (2008). *Struggling to Define a Nation: American Music and the Twentieth Century*, p.54. ISBN 9780520254862. Shown in common time and then in cut time with tied sixteenth & eighth note rather than rest.

[33] Sublette, Ned (2007), *Cuba and Its Music: From the First Drums to the Mambo*, p.134. ISBN 9781556526329. Shown with tied sixteenth & eighth note rather than rest.

[34] Palmer, Robert (1981: 39). *Deep Blues*.

[35] Palmer (1981: 38).

[36] http://www.youtube.com/watch?v=KmmTMg3e5Uo

[37] Kubik, Gerhard (1999: 52). *Africa and the Blues*. Jackson, MI: University Press of Mississippi.

[38] Palmer (1981: 39).

[39] http://www.youtube.com/watch?v=m6mRdPP6wRo&feature=results_video&playnext=1&list=PL84D57DCEA94772EE

[40] Wynton Marsalis states that tresillo is the New Orleans "clave." "Wynton Marsalis part 2." *60 Minutes*. CBS News (Jun 26, 2011).

[41] Schuller, Gunther (1968: 19) *Early Jazz; Its Roots and Musical Development*. New York: Oxford Press.

[42] "[Afro]-Latin rhythms have been absorbed into black American styles far more consistently than into white popular music, despite Latin music's popularity among whites" (Roberts 1979: 41).

[43] Manuel, Peter (2009: 67). *Creolizing Contradance in the Caribbean*. Philadelphia: Temple University Press.

[44] Manuel, Peter (2009: 69). *Creolizing Contradance in the Caribbean*. Philadelphia: Temple University Press.

[45] Acosta, Leonardo (2003: 5). *Cubano Be Cubano Bop; One Hundred Years of Jazz in Cuba*. Washington D.C.: Smithsonian Books.

[46] Mauleón (1999: 4) *Salsa Guidebook for Piano and Ensemble*. Petaluma, California: Sher Music. ISBN 0-9614701-9-4.

[47] Peñalosa, David (2010: 42). *The Clave Matrix; Afro-Cuban Rhythm: Its Principles and African Origins*. Redway, CA: Bembe Inc. ISBN 1-886502-80-3.

[48] Roberts, John Storm (1999: 12) *Latin Jazz*. New York: Schirmer Books.

[49] Sublette, Ned (2008: 125). *The World that made New Orleans: from Spanish silver to Congo Square*. Chicago: Lawrence Hill Books. ISBN 1-55652-958-9

[50] Sublette, Ned (2008:125). *Cuba and its Music; From the First Drums to the Mambo*. Chicago: Chicago Review Press.

[51] Roberts, John Storm (1999: 16) *Latin Jazz*. New York: Schirmer Books.

[52] "Wynton Marsalis part 2." *60 Minutes*. CBS News (Jun 26, 2011).

[53] Morton, "Jelly Roll" (1938: Library of Congress Recording) *The Complete Recordings By Alan Lomax*.

[58] Tanner, Paul, David W Megill, and Maurice Gerow. Jazz. 11th ed. Boston: McGraw-Hill Higher Education, 2009. pgs. 328-331

[59] Matthiesen, Bill (2008: 8). *Habaneras, Maxixies & Tangos The Syncopated Piano Music of Latin America*. Mel Bay. ISBN 0786676353

[60] Sublette, Ned (2008:155). *Cuba and its Music; From the First Drums to the Mambo*. Chicago: Chicago Review Press.

[61] Roberts, John Storm (1999: 40). *The Latin Tinge*. Oxford University Press.

[62] Kunzler's dictionary of Jazz provides two separate entries: blues, an originally African-American genre (p. 128), and the blues form, a widespread musical form (p. 131).

[64] Kubik, Gerhard (1999: 96).

[65] Palmer (1981: 42).

[66] Palmer (1981: 46).

[67] Schuller (1968: 66, 145n.).

[69] Marsalis, Wynton (2000: DVD n.1). *Jazz*. PBS (http://www.pbs.org/jazz/classroom/rhythmicinnovations.htm)

[70] " Jazz and Math: Rhythmic Innovations (http://www.pbs.org/jazz/classroom/rhythmicinnovations.htm)", *PBS.org*. The Wikipedia example shown in half time compared to the source.

[72] Roberts, John Storm 1979. *The Latin Tinge: the impact of Latin American music on the United States*. Oxford.

[73] In 1938 Morton made a series of recordings for the Library of Congress, in which he demonstrated the difference between the two styles.

[74] Gridley, Mark C. (2000: 61). *Jazz Styles: History and Analysis*, 7th ed.

[75] Schuller (1968: 6).

[76] *The New Harvard Dictionary of Music* (1986: 818).

[77] Peñalosa, David (2010: 229). *The Clave Matrix; Afro-Cuban Rhythm: Its Principles and African Origins*. Redway, CA: Bembe Inc. ISBN 1-886502-80-3.

[78] Illustrated well in HBO's program, *Treme*, which has succeeded in researching the jazz culture of New Orleans.

[79] Gridley, Mark C. (2000: 72–73). *Jazz Styles: History and Analysis*, 7th ed.

[88] Handy, W.C. (1941). *Father of the Blues: An Autobiography* Macmillan Co.

[89] Handy, Father (1941), p. 99

[90] *Father of the Blues: An Autobiography*. by W.C. Handy, edited by Arna Bontemps: foreword by Abbe Niles. Da Capo paperback, New York; Macmillan, (1941) ISBN 0-306-80421-2.

[92] *Father of the Blues: An Autobiography*. by W.C. Handy, edited by Arna Bontemps: foreword by Abbe Niles. Macmillan Company, New York; (1941) pages 99, 100. no ISBN in this first printing

[95] Palmer (1968: 67).

[98] Schuller (1968: 91)

[99] Schuller (1968: 93)

[102] Schuller (1968: 88)

[103] See lengthy interviews with Hines in [Nairn] *Earl "Fatha" Hines*: (https://vimeo.com/58414566) – see External Links below.

[106] writes "He tried to avoid the word 'jazz' preferring 'Negro' or 'American' music. He claimed there were only two types of music, 'good' and 'bad' ... And he embraced a phrase coined by his colleague Billy Strayhorn – 'beyond category' – as a liberating principle."

[108] Gunther Schuller Nov 14, 1972. Dance, p 290

[109] Dance p. 260

[110] Floyd, Samuel A., Jr. (1995). *The power of black music: Interpreting its history from Africa to the United States*. New York: Oxford University Press.

[111] Levine, Mark (1995: 171). *The Jazz Theory Book*. Sher Music. ISBN 1883217040

[112] Joachim Berendt. "The Jazz Book". 1981. Page 15.

[113] Charlie Parker quoted by Gerhard Kubik (2005). "Bebop: a case in point. The African Matrix in Jazz Harmonic Practices." (Critical essay) *Black Music Research Journal* 22 Mar. Digital.

[114] Gerhard Kubik (2005). "Bebop: A Case in Point. The African Matrix in Jazz Harmonic Practices." (Critical essay) *Black Music Research Journal* Mar 22, Digital.

[115] Kubik (2005).

[116] Joachim Berendt. *The Jazz Book*. 1981. Page 16.

[117] In 1992 Bauza recorded "Tanga" in the expanded form of an Afro-Cuban suite, consisting of five movements. *Mario Bauza and his Afro-Cuban Orchestra*. Messidor CD (1992).

[118] Peñalosa (2010: 56).

[119] Peñalosa (2010: 131–136).

[120] Bobby Sanabria, posting to the *Latinjazz discussion list* (2008). http://launch.groups.yahoo.com/group/latinjazz/

[121] Dizzy Gillespie, from his book *To Be or Not to Bop* (1985); cited by John Storm Roberts in *Latin Jazz* 1999. p. 77.

[122] "Afro Blue", *Afro Roots* (Mongo Santamaria) Prestige CD 24018-2 (1959).

[123] Peñalosa, David (2010). *The Clave Matrix; Afro-Cuban Rhythm: Its Principles and African Origins* p. 26. Redway, CA: Bembe Inc. ISBN 1-886502-80-3.

[124] Collier, 1978

[126] Levine, Mark (1995: 30). *The Jazz Theory Book*. Sher Music. ISBN 1883217040

[127] Davis, Miles (1989: 234). *The Autobiography*. New York: Touchstone.

[128] After Mark Levine (1995: 29).

[129] http://www.youtube.com/watch?v=DEC8nqT6Rrk

[131] Levine (1995: 30).

[132] http://www.youtube.com/watch?v=me7P9qqBgwI

[133] http://www.youtube.com/watch?v=FEgMsEKICqc

[134] Joachim Berendt. *The Jazz Book*. 1981. Page 21.

[135] Gridley, Mark C. (2000: 444). *Jazz Styles: History and Analysis*, 7th ed.

[136] Tjader, Cal (1959). *Monterey Concerts*. Prestige CD. ASIN: B000000ZCY.

[137] Andy Gonzalez interviewed by Larry Birnbaum. Ed. Boggs, Vernon W. (1992: 297–298). *Salsiology; Afro-Cuban Music and the Evolution of Salsa in New York City*. New York: Greenwood Press. ISBN 0313284687

[139] Allmusic Biography (http://www.allmusic.com/artist/p6300/biography)

[141] "Footprints" *Miles Smiles* (Miles Davis). Columbia CD (1967).

[142] An ancient west central Sudanic stratum of *pentatonic* song composition, often associated with simple work rhythms in a regular meter, but with notable off-beat accents ... reaches back perhaps thousands of years to early West African sorgum agriculturalists—Kubik, Gerhard (1999: 95). *Africa and the Blues*. Jackson, MI: University Press of Mississippi.

[143] Gridley, Mark C. (2000: 270). *Jazz Styles: History and Analysis*, 7th ed.

[144] Map showing distribution of harmony in Africa. Jones, A.M. (1959). *Studies in African Music*. Oxford Press.

[145] After Mark Levin (1995: 235). *The Jazz Theory Book*. Sher Music. ISBN 1883217040

[146] Levine, Mark (1989: 127). *The Jazz Piano Book*. Petaluma, CA: Sher Music. ASIN: B004532DEE

[147] Levine (1989: 127).

[148] After Mark Levine (1989: 127). *The Jazz Piano Book*.

[149] Bair, Jeff (2003: 5). *Cyclic Patterns in John Coltrane's Melodic Vocabulary as Influenced by Nicolas Slonimsky's Thesaurus of Scales and Melodic Patterns: An Analysis of Selected Improvisations*. PhD Thesis. University of North Texas. Web. http://digital.library.unt.edu/ark:/67531/metadc4348/m2/1/high_res_d/dissertation.pdf

[150] Levine, Mark (1995: 205). *The Jazz Theory Book*. Sher Music. ISBN 1883217040

[152] Davis, Miles, with Quincy Troupe (1989: 298) *The Autobiography*. New York: Simon and Schuster.

[153] Dan, Morgenstern (1971). *Down Beat* May 13.

[155] Moore, Kevin (2007) "History and Discography of Irakere". *Timba.com*. (http://www.timba.com/encyclopedia_pages/history-and-discography-of-irakere)

[156] HR-57 Center HR-57 Center for the Preservation of Jazz and Blues, with the six-point mandate. (http://www.hr57.org/hconres57.html)

[157] http://www.jazz.com/jazz-blog/2008/2/27/where-did-our-revolution-go-part-three

[161] Dave Lang, *Perfect Sound Forever*, February 1999. (http://www.furious.com/Perfect/popgroup.html) Access date: November 15, 2008.

[162] Bangs, Lester. "Free Jazz / Punk Rock". *Musician Magazine*, 1979. (http://www.notbored.org/bangs.html) Access date: July 20, 2008.

[165] "... funky, hip music of fearsome technical proficiency ..." (Ian Carr, Digby Fairweather, Brian Priestley, *The Rough Guide Jazz*, 2nd, 2000, "M-Base", p. 500)

[166] "... circular and highly complex polymetric patterns which preserve their danceable character of popular Funk-rhythms despite their internal complexity and asymmetries ..." (Musicologist and musician Ekkehard Jost, Sozialgeschichte des Jazz, 2003, p. 377)

[167] (http://www.allaboutjazz.com/php/musician.php?id=5820) Steve Coleman#Biography

[169] Pianist Vijay Iyer (who was chosen as "Jazz musician of the year 2010" by the Jazz Journalists Association) said: "It's hard to overstate Steve (Coleman's) influence. He's affected more than one generation, as much as anyone since John Coltrane." ((http://online.wsj.com/article/SB10001424052748703302604575294532527380178.html))

[170] "His recombinant ideas about rhythm and form and his eagerness to mentor musicians and build a new vernacular have had a profound effect on American jazz." (Ben Ratliff, (http://www.nytimes.com/2010/06/15/arts/music/15undead.html?_r=1))

[171] Vijay Iyer: "It's not just that you can connect the dots by playing seven or 11 beats. What sits behind his influence is this global perspective on music and life. He has a point of view of what he does and why he does it." ((http://online.wsj.com/article/SB10001424052748703302604575294532527380178.html))

References

- Adorno, Theodor. *Prisms* The MIT Press: Cambridge, MA. 1967.
- Allen, William Francis, Charles Pickard Ware, and Lucy McLim Garrison, eds. 1867. *Slave Songs of the United States*. New York: A Simpson & Co. Electronic edition, Chapel Hill, N. C.: Academic Affairs Library, University of North Carolina at Chapel Hill, 2000 (http://docsouth.unc.edu/church/allen/menu.html).
- Joachim Ernst Berendt, Günther Huesmann (Bearb.): *Das Jazzbuch*. 7. Auflage. S. Fischer Verlag, Frankfurt am Main 2005, ISBN 3-10-003802-9
- Burns, Ken, and Geoffrey C. Ward. 2000. *Jazz—A History of America's Music*. New York: Alfred A. Knopf. Also: The Jazz Film Project, Inc.
- .
- Carr, Ian. *Music Outside: Contemporary Jazz in Britain.* 2nd edition. London: Northway. ISBN 978-0-9550908-6-8
- Collier, James Lincoln. *The Making of Jazz: A Comprehensive History* (Dell Publishing Co., 1978)
- Dance, Stanley (1983). *The World of Earl Hines*. [Includes a 120-page interview with Hines plus many photos]. Da Capo Press. ISBN 0-306-80182-5
- Davis, Miles.
- Downbeat (2009). *The Great Jazz Interviews*: ed Frank Alkyer & Ed Enright. Hal Leonard Books. ISBN 978-1-4234-6384-9
- Elsdon, Peter. 2003. "*The Cambridge Companion to Jazz*, Edited by Mervyn Cooke and David Horn, Cambridge: Cambridge University Press, 2002. Review." *Frankfürter Zeitschrift für Musikwissenschaft* 6:159–75.
- Gang Starr. 2006. *Mass Appeal: The Best of Gang Starr*. CD recording 72435-96708-2-9. New York: Virgin Records.
- Giddins, Gary. 1998. *Visions of Jazz: The First Century* New York: Oxford University Press. ISBN 0-19-507675-3
- Godbolt, Jim. 2005. *A History of Jazz in Britain 1919–50* London: Northway. ISBN 0-9537040-5-X
- Gridley, Mark C. 2004. *Concise Guide to Jazz*, fourth edition. Upper Saddle River, NJ: Pearson/Prentice Hall. ISBN 0-13-182657-3
- Kenney, William Howland. 1993. *Chicago Jazz: A Cultural History, 1904–1930*. New York: Oxford University Press. ISBN 0-19-506453-4 (cloth); paperback reprint 1994 ISBN 0-19-509260-0
- .
- Mandel, Howard. 2007. *Miles, Ornette, Cecil: Jazz Beyond Jazz*. Routledge. ISBN 0-415-96714-7.
- Nairn, Charlie. 1975. *Earl 'Fatha' HInes*: 1 hour 'solo' documentary made in "Blues Alley" Jazz Club, Washington DC, for ATV, England, 1975: produced/directed by Charlie Nairn: original 16mm film plus out-takes of additional tunes from that film archived in British Film Institute Library at bfi.org.uk and http://www.itvstudios.com: DVD copies with Jean Gray Hargrove Music Library [who hold the The Earl Hines Collection/Archive], University of California, Berkeley: also University of Chicago, Hogan Jazz Archive Tulane University New Orleans and Louis Armstrong House Museum Libraries
- Peñalosa, David. 2010. *The Clave Matrix; Afro-Cuban Rhythm: Its Principles and African Origins*. Redway, CA: Bembe Inc. ISBN 1-886502-80-3.
- Porter, Eric. 2002. *What Is This Thing Called Jazz? African American Musicians as Artists, Critics and Activists*. University of California Press, Ltd. London, England.
- Ratliffe, Ben. 2002. *Jazz: A Critic's Guide to the 100 Most Important Recordings*. The New York Times Essential Library. New York: Times Books. ISBN 0-8050-7068-0
- Scaruffi, Piero: *A History of Jazz Music 1900–2000*. 2007. Omniware. ISBN 978-0-9765531-3-7
- Schuller, Gunther. 1968. *Early Jazz: Its Roots and Musical Development.* Oxford University Press. New printing 1986.
- Schuller, Gunther. 1991. *The Swing Era: The Development of Jazz, 1930–1945*. Oxford University Press.

- Searle, Chris. 2008. *Forward Groove: Jazz and the Real World from Louis Armstrong to Gilad Atzmon.* London: Northway. ISBN 978-0-9550908-7-5
- Szwed, John Francis. 2000. *Jazz 101: A Complete Guide to Learning and Loving Jazz.* New York: Hyperion. ISBN 0-7868-8496-7
- Vacher, Peter. 2004. *Soloists and Sidemen: American Jazz Stories.* London: Northway. ISBN 978-0-9537040-4-0
- Yanow, Scott. 2004. *Jazz on Film: The Complete Story of the Musicians and Music Onscreen.* (Backbeat Books) ISBN 0-87930-783-8

Further Reading

- Lyons, Len. 1980. *The 101 Best Jazz Albums: a History of Jazz on Records.* New York: W. Morrow & Co. 476 p., ill. with b&w photos. ISBN 0-688-08720-3 pbk

External links

- Jazz Foundation of America (http://www.jazzfoundation.org/)
- Jazz at the Smithsonian Museum (http://www.smithsonianjazz.org/)
- Encyclopedia of Jazz Musicians (http://www.jazz.com/encyclopedia/index/An)
- Alabama Jazz Hall of Fame website (http://www.jazzhall.com/)
- Jazz Artist and Discography Resource (http://www.allmusic.com/)
- Red Hot Jazz.com (http://www.redhotjazz.com/)
- Jazz at Lincoln Center website (http://www.jalc.org/)
  - Jazz At Lincoln Center Hall of Fame (http://www.jalc.org/halloffame)
- American Jazz Museum (http://www.americanjazzmuseum.com/) website
- The International Archives for the Jazz Organ (http://www.iajo.org/)
- Classic and Contemporary Jazz Music (http://jazzinternetradio.myradiotoolbar.com/)
- The Jazz Archive at Duke University (http://library.duke.edu/specialcollections/collections/jazzindex.html)
- Jazz Festivals in Europe (http://www.jazzfests.net/)
- Free 1920s Jazz Collection (http://www.archive.org/details/Free_20s_Jazz_Collection) available for downloading at Archive.org
- A List of Jazz Lists (http://home.austarnet.com.au/petersykes/jazz100/jazzlinks.html)
- DownBeat's Jazz 101 A Guide to the Music (http://www.downbeat.com/default.asp?sect=education&subsect=jazz) This section of the Downbeat magazine website has several short pages to allow the beginning student of jazz to acquire an education.
- Nairn, Charlie, (1975): *Earl "Fatha" Hines*: (https://vimeo.com/58414566). 1hr documentary filmed at Blues Alley jazz club, Washington DC. Produced and directed by Charlie Nairn for UK ATV Television, 1975. Original 16mm film, plus out-takes of additional tunes, archived in British Film Institute Library at bfi.org.uk; also at http://www.itvstudios.com; DVD copies with the "Jean Gray Hargrove Music Library" (which holds the The Earl Hines Collection/Archive), University of California, Berkeley, California; also at University of Chicago "Hogan Jazz Archive", Tulane University New Orleans and at the Louis Armstrong House Museum Libraries: see also www.jazzonfilm.com/documentaries.
- Jazz (http://topics.nytimes.com/top/reference/timestopics/subjects/j/jazz/index.html) collected news and commentary at *The New York Times*
- Jazz (http://www.theguardian.com/music/jazz) collected news and commentary at *The Guardian*
- Jazz (http://www.dmoz.org/Arts/Music/Styles/J/Jazz/) at the Open Directory Project

David_Temple

David Temple is a British conductor and musical director of Crouch End Festival Chorus[] which he has conducted at the Barbican Centre, Royal Festival Hall and Symphony Hall, Birmingham. He also appears regularly with Hertfordshire Chorus at venues including the Royal Albert Hall, the Barbican, The Sage Gateshead and St Albans Cathedral.

He came to London in 1972 and joined the London Philharmonic Choir where he sang as a tenor under the chorus master John Alldis, performing with conductors including Boult, Stokowski, Solti, Haitink, Tennstedt, Barenboim, Giulini and Rattle. In 1984 he began his work with the newly formed Crouch End Festival Chorus, now a choir of international standing.

As chorus master of the Norfolk and Norwich Festival, he worked closely with Ray Davies of The Kinks on his choral work *The Flatlands*, and collaborated with him on live performances in 2007, 2008, 2009, 2010, 2011 including a performance of The Village Green Preservation Society at the Royal Festival Hall with the London Philharmonic Orchestra, Crouch End Festival Chorus and Ray Davies plus band.

References

Crouch End Festival Chorus is one of the world's leading symphonic choirs, a result achieved under the exclusive direction of David Temple, in his roles as conductor and chorus master. His imaginative and innovative programme planning, including several ambitious commissions, has brought the choir to the forefront of music-making in the UK.

David came to London in 1972 and joined the London Philharmonic Choir, where he sang as a tenor under chorus master John Alldis. He performed with many great conductors including Boult, Stokowski, Solti, Haitink, Tennstedt, Barenboim, Giulini and Rattle.

In 1984 he became musical director of the newly formed Crouch End Festival Chorus. Since then, CEFC has progressed from being a reputable local choral society to a choir of international standing, regularly drawing in substantial audiences at major London venues. David's recent work with the choir includes the Berlioz Te Deum at the 2009 BBC Proms and Britten's War Requiem at the Barbican.

David's interests include pop and rock music, and he has worked with the rock band Travis, as well as with The Divine Comedy and Goldfrapp. David worked closely with Ray Davies of the Kinks on his choral piece The Flatlands, and singers from CEFC accompanied Ray and his band in the climactic concert of the BBC's 2007 Electric Proms - followed by two performances at the Hampton Court Festival. David has now recorded an album of new choral versions of Kinks classics and has recently toured the USA with Ray Davies. The success of the choir's work with Ennio Morricone led to an invitation to accompany superband Oasis in their concert at the 2008 Electric Proms.

Under David's direction, CEFC singers have been invited several times to record background music for the BBC's Doctor Who, and the choir has appeared on the soundtracks of numerous films, including the Hollywood blockbusters Journey to the Center of the Earth and Prince Caspian. David has worked regularly with film composers Gabriel Yared and Ennio Morricone, who is a patron of Crouch End Festival Chorus. Recent recordings with Crouch End Festival Chorus include Noel Gallagher, High Flying Birds: Ray Davies, The Kinks Choral Collection: Katherine Jenkins, Daydream. he has also directed the chorus on the soundtrack to the movie 'The Awakening' starring Rebecca Hall and Dominic West.

David has also been Musical Director of Hertfordshire Chorus since 2000. Hertfordshire Chorus's latest commission is a work by Will Todd, Ode to a Nightingale, which premiered at the Barbican in May 2011.

External links

- David Temple profile, Crouch End Festival Chorus (http://www.cefc.org.uk/musical-director/)

Buxton_Festival

The **Buxton Festival** is an annual summer festival of opera, music, and (since 2000) a literary series, held in Buxton, Derbyshire in England since it began in July 1979.

Buxton Opera House

History

The origins of the Festival date back to 1936 when an annual drama festival was held until 1942 in conjunction with the London-based Old Vic Theatre Company. During the early 1970s, it was best known as one of the UK's most prominent rock festivals, with most major rock bands of the day appearing, including Mott The Hoople, The Faces, Lindisfarne, Canned Heat, Chuck Berry, Nazareth, Edgar Broughton Band, Groundhogs, Sensational Alex Harvey Band, Medicine Head, Brewers Droop, Roy Wood and Wizzard. During July 2007, it was the subject of several features on Jeff Cooper's 'Cooper Collection' show on 106.6 Smooth Radio, whose transmission area includes much of Derbyshire.

Inspired to encourage the restoration of the Buxton Opera House, a classic Frank Matcham building, much respected conductor Anthony Hose (then Head of Music at Welsh National Opera) and Malcolm Fraser (then lecturing in opera at the Royal Northern College of Music in Manchester) saw its potential as a venue for an opera festival. They spent three years planning the first one while restoration was underway. The restored Buxton Opera House was the venue for the first Buxton Festival in 1979 with presentations of *Lucia di Lammermoor* (in its first ever complete performance in Britain), followed by Peter Maxwell Davies' *The Two Fiddlers*.

Productions and performers

From these first successes, the Festival has made a significant impact on the operatic culture of Britain with new productions of rarely-performed operas (such as Britten's *Let's Make an Opera* (1980); Domenico Cimarosa's *Il matrimonio segreto* (1981 & 1993); Kodály's *Háry János* (in its British stage premiere in 1982); Vivaldi's *Griselda* (1983, but not seen anywhere since its original Venice presentation in 1735); Cherubini's Médée (1984, in its original French dialogue never seen in Britain); and, from 1986, many productions of Handel's operas, as well as many others by Cimarosa (in 1989 it presented three).

Performers of the quality of Thomas Allen, Rosalind Plowright, Jean Rigby, Alan Opie, Nigel Kennedy, Cleo Laine, John Ogdon, Alan Bates, Dame Janet Baker, Victoria de los Ángeles,Dame Margaret Price and Sarah Brightman. The current resident orchestra at the festival (as of 2008) is the Northern Chamber Orchestra.

Since 2002, the Festival has presented five or six operas each summer. In 2006 it presented eight operas ranging from Mozart's early *Apollo and Hyacinth* to Dmitri Shostakovich's *The Nose*, but including works by Telemann, Monteverdi, Gluck, Britten and Bizet.

The Buxton Festival Fringe

The Buxton Festival Fringe is an annual open arts festival running at approximately the same time as the Buxton Festival (2013 dates are July 3-21). The festival hosts comedy, theatre, dance, music, street performances, film, performance art, talks and shows for children as well as other impromptu events. The festival celebrated its 30th anniversary in 2009, and that year boasted a record 140 entries with over 500 individual performances. The number of entries has been gradually rising with some 170 in 2013 adding up to nearly 600 individual performances. This makes it one of the largest fully independent Fringes in the United Kingdom, along with Brighton Festival Fringe and Edinburgh Fringe. As well using around fifty venues around the town and outside it, Fringe 2013 includes the professionally managed venue group at Underground Venues [1]. Since 2011 Underground Venues has been programming events at the newly refurbished Pavilion Arts Centre and Studio with Ed Reardon, Isy Suttie and Henning Wehn being among the artists who have featured.

The festival is followed by the International Gilbert & Sullivan Festival, which runs for three weeks in August each year.

See also

- List of opera festivals
- Buxton Opera House [2]

External links

- Arts Derbyshire - Buxton Festival Fringe [3]
- Buxton Festival website [4]
- Buxton Fringe official website [5]

References

[1] http://www.undergroundvenues.co.uk
[2] http://www.buxtonoperahouse.org.uk/festivals/buxton-festival-fringe
[3] http://www.artsderbyshire.org.uk/find_artist/search/artistprofile.asp?ArtistID=334
[4] http://www.buxtonfestival.co.uk/
[5] http://www.buxtonfringe.org.uk

Article Sources and Contributors

Will_Todd *Source*: http://en.wikipedia.org/w/index.php?title=Will_Todd *Contributors*: 2620:0:2E80:A04C:226:BBFF:FE1C:5530, Arctofire, Bender235, Biruitorul, Daniel Rollison, Dur1970, Feline Hymnic, Halmstad, LouieNet, MMetro, MikeConroy, Patchy1, SwisterTwister, Thoughtfortheday, Waacstats, Woohookitty, 10 anonymous edits

Classical_music *Source*: http://en.wikipedia.org/w/index.php?title=Classical_music *Contributors*: 09emvi, 151grand, 1966batfan, 19maxx, 213.253.39.xxx, 21655, 28bytes, 2D, 4meter4, 61.9.128.xxx, 64.228.44.xxx, A3RO, ABF, AS, Acps110, Adam McMaster, Addihockey10, Aeternus, Airconswitch, Aitias, Ajcheema, Akulo, Al.Neuland, Alan Parmenter, Alansohn, Alchemist Jack, Aldoperani, Ale jrb, Alex.muller, AlexChao, Alexius08, AlexiusHoratius, Alfie66, Alfietucker, Alfio, Alksub, Alpha Ursae Minoris, Alxeedo, Amalas, Amazingmate, Amazzing5, Amberglow, Anchoress, Andonic, Andre Engels, Andreworkney, Andrija.b, Andy M. Wang, Andyt922, Angela, Anger22, Antandrus, Apeloverage, AprilSKelly, ArchonMagnus, Ardonik, Arkivmd, Armatura, Artistopia, Ascidian, Astatine211, Aszazin, Avoided, Bagatelle, BarretB, Basketball3r22, Battem, Beeyan, Before My Ken, Ben Tibbetts, Benson85, Benthelog, Bevo74, Bfigura, BigMak2027, Bill edmond, Bill-on-the-Hill, Billy4, Binksternet, Binyamin Goldstein, Bleh fu, Blehfu, Blethering Scot, Bluewave, Bobertbobertboert, Bobnorwal, Bobo192, Bongwarrior, Boulaur, Brambleclawx, Brandon, Bryan Derksen, Bsadowski1, Caltas, CambridgeBayWeather, Camembert, Can't sleep, clown will eat me, CanadianLinuxUser, Capricorn42, Captain Noobtastic, Carmichael, Carol wg, Caseywestergaard, Cessator, Cgs, Chaos, Charlene chan, Charles Talbut Onions, Chris K, Chris Roy, Chub-chub008, Chun-hian, Cielomobile, Citicat, Civil Engineer III, ClarinetistJuan, Classiko, ClaudineChionh, Cleared as filed, Cmichael14, Coemgenus, Cometstyles, Contrafagotto, Conversion script, Cooooler, Cosmic Latte, CosmicJake, Cosprings, Cpe1704tks, Crculver, Crochet, Cyrusfusion1, DCGeist, DVdm, DabMachine, Danmuz, DannyDaWriter, Darko4t, Darth Panda, Darthgriz98, Dave.Dunford, Davemck, DavidRF, Dbiel, Dcooper, Deadworm222, Deb, Deepanjan nag, Deniz yildiz, DerHexer, DerechoReguerraz, Deskford, Dgillett, Dicklyon, Discospinster, Dittyh39, Doc Tropics, DocWatson42, Dogru144, Donner60, Doppio bursitis, Download, Dpbsmith, DrG, DrHok, DragonflySixtyseven, Drak2, Dratman, Drdr1989, Dreamyshade, Dtgm, Dysprosia, E tac, Easydude, Edward321, Efficiencyjacky154, Elliotreed, Enigma7seven, Ensign beedrill, Epbr123, Epolk, Erik Zachte, Euchiasmus, Evercat, Evil saltine, Ewangowan, Excirial, Fablau, Fattums, Favonian, Felix Folio Secundus, Fieldday-sunday, Flewis, Florian Blaschke, Flosculus, Formulax, Francis Schonken, FreeKnowledgeCreator, Frosty0814snowman, Fæ, GFHandel, GRAHAMUK, GabeMc, Gadfium, Gaff, Gail, Gandalf1491, Garret Beaumain, Gary King, Gayane Gharakhanyan, Gfradsp, Gheuf, Giftlite, Gilgamesh he, Ginstar, GladstoneBrookes, Glen, Glenn, Gogo Dodo, Graham87, Grammarxxx, GreatWhiteNortherner, GripTheHusk, Grover cleveland, Gtg204y, Gurch, Guruswara, Guy Peters, Gökhan, Hajatvrc, Haldraper, HalfShadow, Hapless Hero, Harry Doodli, HaydnAustria, Haymouse, Hebrides, Henage, Henry McClean, Herschelkrustofsky, HexaChord, Heya444, HiLo48, Hmains, Human.v2.0, Hyacinth, ILike2BeAnonymous, ILoveEditingmE, Iammaeai, Ignatzmice, Igoldste, Imic, InfoCan, InnocuousPseudonym, Insanity Incarnate, Iokseng, Isnow, It Is Me Here, Ixfd64, J-klem, J.delanoy, JBanks94, JBomb018, JBowman79, JForget, JNW, JRHorse, Jac16888, Jacj, Jackelfive, Jafeluv, Jahsonic, Jakew, JamesBurns, Jamesx12345, Jarrod Lachlan Dennis, Jauerback, Javierito92, Jaybirdjustino77, Jeff G., Jeffrey Mall, JesseHogan, Jfucinos, Jim1138, Jj137, Joefromrandb, John Rocher, John of Reading, JohnnyB256, Johtty, Jojhutton, Joker9797, Jonathan.s.kt, Jotsko, Jounisakari, Jp307803, Jpbowen, Jpeob, Jpo, Jshantier, Jubileeclipman, Juliancolton, Julie Martello, Jusdafax, KDTashi, Ka Faraq Gatri, Kannie, Karada, Karmafist, Kaysov, Kchishol1970, Keenan Pepper, Keepssouth, Keithcadenza, Keithg, Kerriforwiki, Kerrio, Khalid Mahmood, Khazar2, Kimtaro, Khonnichiwa, Kleinzach, KlingsorOfOz, KnowledgeOfSelf, KoshVorlon, Kpjas, Kri, Kusma, Kwj2772, LAX, Lament, Lawrence H K, LazyStarryNights, Ld100, Leszek Jańczuk, Lexor, Lighthead, Limepink bubbles, Little Mountain 5, Llamaah, Logan, Lololololololololololo8, Louisshomar, Luckybeargod, Lugia2453, Luna Santin, MER-C, MLKLewis, Madder, Madhero88, Madinalake25, Magicpiano, Major A-flat, Makecat, Makemi, Malbi, Mallocks, Malo, Mandarax, Manuel Anastácio, Mar4d, Marcus2, Marek69, MarkBolton, MarkBuckles, MarsRover, Martin451, Marzie1991, Masahiko, MatrixFrog, Matt.kaner, Max Naylor, Maxis ftw, Mayalld, Mayooranathan, Meeror, Meet2kunal, Mel Etitis, Melsaran, Mendaliv, Merefoix, Merphant, Metsramsknicks, Michael Bednarek, Michael Furey, Michael Hardy, Michael Snow, Michaelrayw2, Michaelsayers, Mike Rosoft, Mindspillage, Minimac, Mirandamir, Mishuletz, Missmarple, Mitchbaker97, MmmmJoel, Modernist, Moostyrio, Motaros, Mountainclimber4456, Mr Morden76, MrFizyx, MusicMaker5376, MusicTree3, Musicindia1, Musiphil, MyFavoriteMartin, Myopic Bookworm, Naddy, Nadiarehman, NameIsRon, Narutolovehinata5, Natural Cut, Naxosb2c, Nayra, Nayvik, NeilN, Neiler, Nevilley, Nietzsche 2, Nikkimaria, Noir, Nposs, Ntsimp, Ohconfucius, Ohnoitsjamie, OldakQuill, Omicronpersei8, Omortimer, OnBeyondZebrax, Onco p53, Opus33, Opus88888, OwenX, PEPSI2K7, PTSE, Patxi lurra, Paul A, Pavel Vozenilek, Pax:Vobiscum, Pcpcpc, Pedro, Penguinsruleman, Penguinsruleman1, Peoplefromarizona, Persian Poet Gal, Peter Fleet, Petrb, Phantomsteve, Philip Trueman, Physicistjedi, Picard1102000, Pigman, Pigstinky, Pinecone, Pjones235, Pladask, Plasticup, Poitypoity, Ponyo, Poorsod, Porntepj, Pouyakhani, Ppntori, PranksterTurtle, Prashanthns, Prof saxx, Prolog, Psy guy, PuzzletChung, Quadell, Quantumobserver, Quercusrobur, QuiteUnusual, RafaAzevedo, Ralphscheider42, Rayford, Rdsmith4, Realrobwestwood, Redheylin, Rettetast, RexNL, Rheostatik, Rich Farmbrough, Rigaudon, Rintrah, Rjwilmsi, Rmhermen, Robert Merkel, RobertG, Robertvan1, Robin klein, Rod Corkin, Rohith goura, Roleplayer, Ronaldo Guevara, Ronhjones, Rossiver, Rothorpe, Rror, Ryulong, Sadalmelik, Sagaci, Sainsburysandy, Salvio giuliano, Sam Korn, Samnotwil, Sanderling, Sangak, Saturn star, Schissel, Scrutchfield, Sdasdasda, Seahorseruler, Semperf, Sentausa, Senthilvelsp, Seunghun, Shauno1863, Sheece, Shimmin, Shreevatsa, Shrillserious, Sietse Snel, Silverxxx, SimonMayer, SimonP, SingCal, Skomorokh, Slambo, Sluzzelin, Smedley Hirkum, Smerus, Snasr, Snow1215, Snowmanmelting, Snowolf, Sobolewski, Sodium, Soler97, Soliloquial, SpNeo, SpaceFlight89, Spagaroony, Sparkygravity, Springnuts, Squandermania, Standardfact, Startswithj, Stephen Burnett, Steven Zhang, StillTrill, Stirling Newberry, Svelyka, Symane, TUF-KAT, Tamfang, Tangent747, Tangotango, Tarquin, Tbhotch, Tedius Zanarukando, Tempodivalse, The Anome, The Rationalist, The Thing That Should Not Be, TheKid, TheLH, TheOldJacobite, Themfromspace, Thepianst, Thingg, Tiberius Curtainsmith, Tide rolls, Tim Shell, TimNelson, TimVickers, Toccata quarta, Toddst1, Tom-, Tommy2010, Tony Fox, Topdragon111, Topfilmi, Treisijs, Trwilliams1, Tty29a, Uncle Dick, Uncool User, UninvitedCompany, Urbanguy1, User2004, Vague Rant, Velella, Versus22, Verticalsearch, Viajero, Violncello, Vipinhari, Vivaldi 1612, Voceditenore, Vrenator, Vvven, WAvegetarian, Wackywace, WadeSimMiser, Wahoofive, WailingWailer, Wangry, Wavelength, Wayne Slam, Wellington, WereSpielChequers, Wetman, Whjayg, Widr, WikHead, Wiki libs, Wikilord321, WikipedianMarlith, Wikipelli, Wikiweev, William Avery, Willking1979, Winkinblinkinnod, Woohookitty, Woppjocky, Wshun, Wurblzap, XTerminator2000, Xp54321, Yamamoto Ichiro, Yidisheryid, Yomomma 94, Yousuckworld, Zazaban, Zenibus, ZooFari, האמסיפה123, וו, בוט, , , 1359 anonymous edits

Composer *Source*: http://en.wikipedia.org/w/index.php?title=Composer *Contributors*: 16@r, 2601:7:7600:36:2819:60E7:D641:AC6B, Acalamari, AdjustShift, Aeternus, Ahoerstemeier, Alansohn, Alcmaeonid, Alexander.stohr, Alison, Alperen, Amaury, Anaya, Andonic, Andrija.b, Anhaoxing, Antandrus, Atavi, Athaenara, AtticusX, AubreyEllenShomo, Avjoska, Axrai, Badagnani, Bbullot, Bobo192, Bomzaiya, Bongwarrior, Borgx, Brian0918, Brion VIBBER, Camembert, Ceoil, Cesar.oliveira, Chun-hian, Circeus, Climax33, Cosprings, Crvenotopce, Cunningofunreason, Cybermonsters, CyclePat, DARTH SIDIOUS 2, DVD R W, Dafoeberezin3494, DannyDaWriter, David Gerard, Davidsilver65, Dbfirs, DearPrudence, Dendodge, Devourer09, Discospinster, Docu, Dzordzm, East718, Epbr123, Eyeresist, Fabriziom, FalconL, Filmerbob, Firstsoundent, Galoubet, Gauss, Gerda Arendt, Glenn, Goaty69, Gregbard, Gurch, Gwernol, Hrdinský, Hrymur, Hyacinth, Indon, Iohannes Animosus, JDOCallaghan, James599, Jan Blanicky, Jauhienij, Jaxl, JdeJ, Jongleur100, Jorcoga, Joyous!, Justme89, Kala, Karol Langner, Keegan, Kelcey s, Khukri, Kimse, Kleinzach, Kungfuadam, LeaveSleaves, LeoNomis, Lightmouse, Los3, Lucius1976, M-rcus, Madder, Malhonen, Mclay1, Megamanx1232, Meladina, Melchoir, Methecooldude, Michael Bednarek, Michael Snow, Michelekaschub, Mormegil, Moverton, Narutorx7777, NerdyScienceDude, Netoholic, Nousernamesleft, Oculi, Ohnoitsjamie, Oscar, Owen, Pacific1982, PamD, Papa November, Paralipsis, Parkjunwung, PatrikR, Paulmcdonald, Pegah, Phantomsteve, Philip Trueman, PhotoBox, Pichpich, Pladask, Prolog, PuzzletChung, QuartierLatin1968, Quinceps, Ranveig, Raven in Orbit, Redheylin, Repas, Rheostatik, Ridm, RobertG, Robin klein, Rohirok, Romanm, Ronaldomundo, Rtay, S Marshall, Sango123, SchreiberBike, Sherool, ShimShem, SimonP, Speed CG, Spyros Pantenas, Stephenb, Stickee, Sunilpatnisbvmusic, SusikMkr, TUF-KAT, Talking, Tamadamnkud, TastyPoutine, Tempodivalse, Theone00, Thorman3, Tibetibet, Tomkurts, Tommy Kronkvist, Treisijs, Trinnan, Underdawg, UninvitedCompany, Urbanguy1, Versus22, Vina, Voceditenore, Wahoofive, Wayward, Whjayg, WikiCantona, WikiPBia, Wknight94, Wolfcm, Xanucia, Yidisheryid, Александър, Анна Волкова, ПешСай, 230 anonymous edits

Pianist *Source*: http://en.wikipedia.org/w/index.php?title=Pianist *Contributors*: "D", 123Apropos, 151.24.191.xxx, 20021996js, 212.211.70.xxx, A More Perfect Onion, A412, A455bcd9, Abeg92, Abstrator, Adeliine, Aeternus, Agaas@yahoo.com, Alexandermassey, AlkanSite, Allenstone, Andonic, Andrew Winston, Andylkl, Antandrus, AquaMarine, ArdeshirBozorg, Artszero, Artvandelay, Asqueella, AtzeKatze, AugPi, Augusta1, Bahar101, Baumannp, BazookaJoe, Bearcat, Bearfoot, Ben Tibbetts, Ben jaminhr, Bigbuks999, Billbike, Bjankuloski06en, Blauer Heinrich, Bluerasberry, Bluezy, Bobbyfred, Bobo192, Bomac, Bordechr000, Brequinda, Brisellirc, ButtuHutagalung, Calton, Camembert, Can't sleep, clown will eat me, CanadianLinuxUser, Champion1000, Chatdelaines, Chillum, Chris 73, Classicalmusicman, Claudioarrau, Clestur, CommonsDelinker, Conversion script, Cricket02, Crisso, Csnewman, Curero, DJArgus, DMacks, Damon Mah, Daniel Mietchen, Danielk212, Darvas, David Martland, Deb, Deflective, Denizz, Devans00, DistinctHead, Dpotter, Dr. Lecter, Drhoehl, Dxvxd, EamonnPKeane, Easterbradford, Ellywa, Empty Buffer, Etincelles, Euthymos, Fanny40, Fanny44, Farsee50, FayssalF, Flamurai, Floris V, Flyback, Francisco BR, Fuhghettaboutit, Gavriliada, GeeJo, Gemanieto, Genb2004, Gieter, Giggy, Gilliam, Giventends, Godowsky, Gogo Dodo, Golbez, Graham87, Groenekikker, Grutness, Gwernol, H2O, HaeB, Havefun2007, Here, Hlwarrior, Hughcharlesparker, ILaw, Idono, Ignaciospain1, Infrogmation, Isnow, Ivanescence, Izikq, JYOuyang, JaGa, Jack burns 26, JackofOz, Janne Uusitalo, Jcw69, Je5s3r, JeromeParr, Jersyko, Jesta, Jfdwolff, Jfurr1981, Jim10701, Jimmothyjimbo, Jjo Cheau Ling, Jlittlet, Jncraton, JohnOwens, JohnWYC, Jowa fan, Jsrwang, Juliancolton, Karenhippo, Ken Gallager, Keresaspa, Keron Cyst, Kevyn, Kleinzach, Klemen Kocjancic, KnowledgeOfSelf, Kolindigo, Koyaanis Qatsi, Kozuch, Kross, Kwamikagami, Kyo765, Ladj, Lady Mondegreen, Larrybob, LeaveSleaves, Leonard Vertighel, Ligart, Lilleplym24, LiniShu, Lisa lee 422, Lpgeffen, Lugia2453, Luna Santin, M A Mason, M ajith, Ma'ame Michu, Macphisto12, Madchester, Magister Mathematicae, Mahahahaneapneap, Malcolmxl5, Malitte, Mandarax, Mandel, Marktreut, MasterEditor7, Materialscientist, Mattbr, Maurice Carbonaro, Mauricette, Mayeast, Mayumashu, Mdf, Meeso, MegX, Mel Etitis, Merphant, Mhare40, Michaelsayers, Mikeo, Mikimilano, Mintguy, Missmarple, MoiraMoira, Mrchops10, MusicLoverGuy, Musichere, NYCPR, Nan, Nancyhoule, Neopoiesis, Nicke Lilltroll, Noblegoose, Nova Weaver, Nycmoomin, Oakshade, Oliver202, Opiaterein, Opus33, Ortolan88, OwenX, Parsifal, Pdxgoat, Pgeffen, Pgldbrg, Phoenix-ashes, Pianette7, Piano12345, Pianojazz, Pianoman199, Pixie2000, Pladask, Pm88, Prhartcom, PrincessWortheverything, Px21, Q8-falcon, Raban, Rafa82, Ralphscheider42, Ranjithsutari, Raptoris, RasputinAXP, Rconroy, Redheylin, Redtube214, Reechard, Reneec, Rentaferret, Rfl, Rich Farmbrough, Rigaudon, RjLesch, Rjstott, Rlove, Robbiewolf, Robin Weng, Royjohansson, Rsholmes, SKMUSIC, Sangita Lal, Scientizzle, Scott5114, Scriberius, Scwlong, Scymso, SheltonSchyler, Shen2000, SimonMayer, Skagedal, Snow steed, Solipsist, Some jerk on the Internet, SoulBrotherNumberOne, Special-T, Specificnotes, Spencer195, Stan Shebs, Steed Asprey - 171, SteveA, Stormie, Struway, Suisui, Superollywood, Sxeptomaniac, Szarka, TBHecht, TakuyaMurata, Tarquin, Tassedethe, Tennesseedude, The Thing That Should Not Be, TheCustomOfLife, TheMidnighters, Timbers, Tintazul, Tobias Bergemann, Tom, Torres1287, Tran Quoc Dung, Trident13, Trontonian, Ubiquity, Urbanguy1, ValentinKostadinov, Valeriahawaii, Viajero, Vicomte de Valmont, Violncello, Viriditas, Vladmirfish, Vxlover, Wallie, Wching88, Wedit, White Mage Cid, Whkoh, Wikipelli, Wikiversion, William Avery, Window, YoramIH, Zeisseng, Znethru, ბოცონტი, 499 anonymous edits

Durham_School *Source*: http://en.wikipedia.org/w/index.php?title=Durham_School *Contributors*: 29 00 pie, Acq3, Acsian88, Ajpg, Angela26, Angusmclellan, Baseball Watcher, Bcp67, Bgwhite, Bleaney, Bobblehead, Boleyn, BrownHairedGirl, CT Cooper, ChipUK, Coyets, David Biddulph, Deathardath, Djln, Doncram, DuncanHill, Duncharris, Fozi999, Freezee90,

FruitMonkey, Gabbe, GoingBatty, GordyB, Gurch, Guroadrunner, Hotcodders, Icairns, J.delanoy, Jeltz, Jenks24, JoanneB, Jonathan Oldenbuck, Jpeob, JustAGal, Kanguole, LilHelpa, Lutionium, Maralia, Margareta850, Mediran, Meters, Mhockey, Myosotis Scorpioides, Neddyseagoon, Nedrutland, Nick Number, Oceaniron07, Oneillben5, PBS, Perebourne, Petrb, Pflastertreter, Philismint, Pit-yacker, Quig, R'n'B, Rgmmortimore, Rich257, Rjwilmsi, Rockstar983, Saga City, SamuelTheGhost, Scipio Carthage, Shploom, Sjakkalle, Smelly227, Sp8328, Struway, Stuffed cat, Suslindisambiguator, Tabletop, Teach46, Ted87, Throbblefoot, Tim!, Tony allen 1, Toon05, TubularWorld, Usgnus, Vyom25, Wavelength, Woohookitty, Wrelwser43, Xn4, Zzuuzz, 227 anonymous edits

University_of_Bristol *Source*: http://en.wikipedia.org/w/index.php?title=University_of_Bristol *Contributors*: (aeropagitica), 1exec1, AJR, AJVKey, AP1787, AZ English, Aannoonn, AdultSwim, AgentCDE, Aled D, Alex Law, Alexbouditsky, Aloneinthewild, Andycjp, Angela, Anstey-p, Aristophanes68, Arjun01, Arpingstone, ArthurGD, Asyndeton, Auntof6, Awiseman, Axt, Az1568, AzrielHex, Barfooz, Beeblebrox87, Behun, Benbristol, Benjaminevans82, Berks105, Birkett, Blake-, Bleaney, Blytonite, Bornintheguz, Boron12, Brabzzz, Brichcja, BristolUni, C.Fred, CT Cooper, Camembert, Can't sleep, clown will eat me, Catapult, Cgs, Chaz62, Chrisjj, Colonies Chris, CommonsDelinker, Compfreak7, Confusedmiked, Crispness, D6, DanielJudeCook, Davidkinnen, DerHexer, DinosaursLoveExistence, Dirac1933, Dmitri Lytov, Dnthomps, Draeco, Dribull, DubaiTerminator, EQuacquarelli, Eddieqs, Edfrankl, Edwardx, Egberts, Elf-friend, EnglishTea4me, Enzino, Erik9, Erp, Erpbridge, Eskandarany, Espresso Addict, Esrever, Etimbo, Evil saltine, FTGHSmith, Feketekave, Firsfron, Flaming Ferrari, Forfica, Francium12, Freedomflag, Fremsley, Frietjes, Gabbe, Gaius Cornelius, Gasheadsteve, Genuine918, George1023, Gobonobo, Ground Zero, Grstain, Gun Powder Ma, Handheldpenguin, Harry R, Harryman100, Herbee, Hiftikhar, Hmains, Hoacucxanh9999, IainHallam, Iainscott, Ian3055, Icairns, Iceaxejuggler, IcyEd, Imran, Intgr, Invisible man, Ipetrow, Itsmejudith, Jameswilson, Jbitel92, Jc1326, Jenblower, Jezhotwells, Jimmychurchill, Jm6852, John Fulham, John K, John of Reading, Johncalhoun21, JuJube, Jusdafax, Just Another Dan, KD Tries Again, KF, Kane5187, Kangaroogirl, Kesal, Kesla, Kevinteacosy, Kgwo1972, King Logo Loves Cake, Koavf, Koyn, Lamppost51, Ldyajmf, Legis, Leszek Jańczuk, Leushenko, LilHelpa, London25, Lucy Stephens, Lupin, Luzonaga, MER-C, MK8, Madanarchist, Magnus Manske, Mahahahaneapneap, Mareklug, Mark t young, Mark83, Martin451, MartinHarper, MartinPoulter, Masonpatriot, Materialscientist, Mattbr, Melchoir, Mgumn, Mhockey, Mholland, Michael Devore, Morwen, MrBeast, Napoleonchowmein, Nate1481, Netoholic, Nevilley, Newtextwikiuser, Nicholasphan, Nick, Nymphicus, Ohconfucius, OldakQuill, PDH, Pauln1964, Pcpcpc, Pgk, Phaedriel, Pjb007, Plustgarten, Pogsurf, Postdlf, Preejohn, Proteinpete, Purple Duke, Qwfp, Rangoon11, Rbrwr, RedCoat10, RedWolf, Reginaldbright, Rich Farmbrough, Richard cocks, Richard temple, Risker, Rjwilmsi, Rnt20, Robina Fox, Robincognito, Rodhullandemu, Rodw, Roisterer, RuiRed, Rwendland, SE7, Saalstin, Sabinelincoln, SamuelTheGhost, Sanaanwar1, Sandrobt, Sars, Scurra, Shanangela, Shoeofdeath, SimonP, Skier Dude, Sky, Slj, Smewn 7321, Sodium, Sohmati, Sp8328, Splash, Spreadjoy, SteinbDJ, Steinsky, Stephen MUFC, Stephenb, Tabletop, Tartanperil, TashaSavvy, Tassedethe, Teatreez, TerriersFan, The Bushranger, The Evil Spartan, The Sage of Stamford, Thebadgermachine, Thedjatclubrock, Theon, Tibetan Prayer, Tim giddings, Tim!, Timrollpickering, Tmc5000, Tobinwebb, Tomsega, Tony1, Torger, Tpbradbury, TubularWorld, UK-Logician-2006, Userabc, Vanish2, Vanpump, Versus22, Vgranucci, Voltarios, Wereon, Widr, Wikiball18, Wikinorthernireland, Wikisamia, Woohookitty, X201, X96lee15, XLerate, Xevious, Xyphoid, Yogurtchutney7, Yunshui, Zhu Haifeng, Zoicon5, Zzuuzz, 636 anonymous edits

Jazz *Source*: http://en.wikipedia.org/w/index.php?title=Jazz *Contributors*: "alyosha", (jarbarf), -Midorihana-, .:Ajvol:., 09aidepikiw, 11dkaelin, 165.121.120.xxx, 1archie99, 1nic, 208.200.242.xxx, 23ollie23, 2602:306:CEF2:D460:ADE7:1CB6:5DBF:74D0, 28421u2232nfenfcenc, 2jpech, 4twenty42o, 5theye, 61.9.128.xxx, 72Dino, 78.26, 78963lololol, 88lesix88, A little insignificant, ABF, ACSE, Aaron Brenneman, Aaron charles, Abce2, Abhishek191288, Acather96, Achowat, Acmkes, Acroterion, Adamshappy, Aduderocker, Aeternus, Afterwriting, Agiorgio84, Ahoerstemeier, Ajeya Tatake, Aktiver Arbeiter, Akuv, Alan W, Alanjohns, Alansohn, Alex.gosselin, Alex3917, Alex43223, AlexR, AlexiusHoratius, Alien life form, AllTalking, Allenstone, AllyD, Alsee, Altenmann, Amcaja, Americasroof, Amherst99, AmitTheSomthing, Analbumface, Andonic, AndreNatas, Andrew Norman, Andrew12342, Andrewman327, Andros 1337, Andux, Andy Dingley, Andytuba, Anger22, Ankit Maity, AnonUser, Anonymous Dissident, Antandrus, Antonimos, Antonis Christofides, Anturiaethwr, Aparker621, Aperson1234567, Aphaia, Apollotrombone, Appraiser, ArglebargleIV, Aristotle1990, Arrrknot, Art LaPella, Arthena, Astatine-210, Atechi, Atif.t2, Atlant, Atomino, Atraxani, Atsirk20, AuburnPilot, Auntof6, Auréola, Austin Hair, Autiger, Avb, Avoided, Awseddeswa, B-Machine, B00P, BNutzer, BTP51, Baa, BarretB, Barrettmagic, Bartledan, BassHistory, Basswulf, Bbq332, Bcummins12321, Bdesham, Beetstra, Bejnar, Beland, Belgar, Belizefan, Ben Ben, Ben Tibbetts, Benlevett, Bentogoa, Bevo, Bgwhite, Bib, Bifftar, BigBang6, BigFatBuddha, BigPimpinBrah, Biker Biker, Binksternet, Blackpower, Blanchardb, Blehfu, BluesJazz56, Blueswalk, Bmusician, Bobo192, Bobshowrocks, Bobynose141, Bongwarrior, Bonsaibanzai, Boredmanbored, BorgQueen, Borisblue, Bouch101, Bped1985, Brabblebrex, Bradfordalderman, Brady Ames 311, Brandmeister, Brandmeister (old), Branman515, Brianreading, Brianshazaaam, Bridon2, BrownHairedGirl, Bsharvy, Btrabaris, Bubblegumski, Burgundavia, Busterb132, Butterchurn, C777, CDutcher, CGBass, CIreland, CKRamiscal, CRICKETLtd, CSWarren, Cab88, Cacycle, Calliopejen1, Caltas, CambridgeBayWeather, Camembert, CamperStrike, Can't sleep, clown will eat me, CanadianLinuxUser, Canterbury Tail, CapitalR, Capricorn42, CardinalDan, Cardshark04, Carlos T. Blackburn, Carol wg, Carptrash, Catapult, Catbar, Catgut, Catslash, Cbjazz, Cckkab, Cdg1072, CecilWard, Celarnor, Ceranthor, Cewvero, Ceyockey, Chaizzilla, CharlotteWebb, Chatter, Chazinc121, CheMechanical, Ched, Cheese man 300, Cherubinirules, ChillDeity, Cholmes75, Chris Capoccia, Chris the speller, ChrisCork, Christian List, Chubbles, Chun-hian, Chykensa, Cielomobile, Circeus, CityOfSilver, Claygreenberg, Codetiger, Coemgenus, Coffee, CommonsDelinker, Comoestassenorita, Computerhater, Confuzion, Conorobradaigh, Conversion script, CoolJazz5, Coolandnerdy777, Cosprings, Courcelles, CpprHd, Creation7689, Crossbow 123, Crosstimer, CryptoDerk, Crystallina, Cxz111, Cynicism addict, D, D-Notice, D6, DARTH SIDIOUS 2, Dabears29, Daddy Yankee, Dakota154, Damicatz, Dan56, DanielRigal, Dar-Ape, DarkMattata, Darth Panda, Darthgriz98, Dave souza, Dave.Dunford, Davepape, David Gerard, Dawdler, Dbabbitt, Dcfunnywalker, Dcooper, Deeceevoice, Deltabeignet, DennyColt, Dentalplans, Deor, Der Golem, DerHexer, Devin122, Diannaa, Dicklyon, Dimadick, DirectEdge, DirectorG, Disco1979, Discographer, Discospinster, Dissolve, Djegan, Djlounge, Dogru144, Dominas, Donner60, Dosowski, Dougmerritt, Dougofborg, DougsTech, Dr clave, Dr. Blofeld, Draicone, Drewster414, DriveMySol, Drmagic, Drmies, Druiloor, Drumsac, DuncanHill, Dureo, Durova, Dusti, DutchmanInDisguise, Dvd-junkie, Dwarf Kirlston, E-Kartoffel, E2eamon, EEPROM Eagle, EJF, Eagle4000, Earlypsychosis, Ebrownless, Edderso, EddieHugh, Edison, Editor437, EdouardCabane, Eeekster, Egmontaz, Ehabkost, Ehheh, Elderpops, Electricladyland69, Elizmarrow, Elonka, EmO4LiF3, Emendo, Emmaloupe, Enchanter, Endorfinized, Entergarch, Enviroboy, Epbr123, Epicurus B., Eplack, Erebus Morgaine, Ericrock, EronMain, Esanchez7587, Espoo, Eulen, Evanreyes, Everyking, EvilStormSam, Exarion1, Excirial, Extransit, Ezn, FACUP87, FF2010, FPSFan123, Fahadsadah, Fair Deal, Fantacular, Faradayplank, Farside, Favonian, FayssalF, Felix Folio Secundus, Ferdinand Pienaar, Figaro-ahp, Fire123456789, Fishbulb, Flatterworld, Flewis, Flint McRae, Fluffernutter, FluffySquid, FlyingToaster, Foobarnix, Fpittui, Francs2000, Fratrep, Freakofnurture, Fred924, Fredil Yupigo, Fredrik, FredrikT, FreeKresge, Freechild, Freekee, Freespazz, FriedMilk, Frymaster, FukinTROLL, Funeral, Funkies, FunkyFly, Funnyfarmofdoom, Furrykef, Futurebird, Fyyer, Fæ, G Clark, GHe, GWFrog, Gabbe, Gabspeck, Gadfium, Gail, Galinskykid, Galwhaa, Gareth E Kegg, Gareth Owen, Gartnait, Garyzx, Garzo, Gautam3, Gavin Wilson, GcSwRhIc, GenOrl, Geneb1955, Generaljim99, Geniac, Geniusboy88, Geoduck75, GeorgeJeffersonton, GeorgeMoney, Gerald Farinas, Gerardhoward, Gettingitright100, Gggggghhh, Ghazer, Gheuf, Ghibson Hudson, Gibson Flying V, Gil Gamesh, Gilliam, Giraffedata, Girlntraining04, Glacialfox, Glassbreaker5791, Goethean, Gogo Dodo, GoingBatty, Gokberkcinbis, Goldeneye232, Gongshow, Good Intentions, GoodBlackCat, Goodnightmush, GorillaWarfare, Gracenotes, GraemeL, Grafen, Grafikm fr, Graham87, Grayshi, Grblomerth, Greatestrowerever, Green moofle, Gregoryoko, GripTheHusk, Gscshoyru, Guerillero, Guntramshatterhand, Gurch, Gurchzilla, Guy M, Gwernol, Gwyndon, Gyrofrog, Habj, Hadal, Hajor, Halaster, Haldraper, HalfUgly, Hallows AG, Haribo 123456789, Hatyy12, Hayabusa future, Haymouse, Hayzerthelayzer, HazelAB, Hdstubbs, Hector1000, Helo123456, Henage, Henrnes, Henry Flower, Heracles31, HermanHiddema, Hewhoamareismyself, HexaChord, Highnotes4ever, HighwayCello, HisSpaceResearch, Hisothiro, Hmains, Hollandave, Hollomis, Hoops gza, Hoot, Hubertfarnsworth, Husond, Hyacinth, Hydrogen Iodide, I dream of horses, II MusLiM HyBRiD II, ILoveFuturama, Ian Pitchford, Iced Kola, Idmillington, Igoldste, Ihasbrainsftw, Ilikeitupthebutt, Ilyushka88, Imanzia, Iminnocent0, Infrogmation, Instinct, Intelati, Intelligentsium, Inter16, Intoronto1125, Irishguy, Isbass, Isidore, Ixfd64, Izno, J.delanoy, J04n, JAZZNAN, JCraw, JFA2009, JForget, JPX7, JaGa, Jacek Kendysz, Jack haze 55, JackSparrow Ninja, Jackfork, Jackk, Jackpinder170901, Jacksinterweb, Jacobfred, Jafeluv, Jagged 85, Jahsonic, Jake Wartenberg, James.beckwith, Jamesontai, Jamiecrocker0, Jamieli, Jan eissfeldt, Janejellyroll, Jarrod Lachlan Dennis, Jawschlech, Jayojay, Jazz N Media, Jazz zangeres Anne-J, Jazz+, Jazzcorner, Jazzmans99, Jazzmusic, Jazztastic, Jcardinal, Jcmo, Jdavidb, Jdmatouk, Je'Ran, Jeandré du Toit, Jeffq, Jeffrey Mall, Jeffrey O. Gustafson, Jengod, Jennifer Parks, Jerome Kohl, Jerry.Ivie, Jesta, Jfdwolff, Jfitzg, Jim Apple, Jim1138, Jimmyawsam, Jimstahl333, Jiy, Jk1lee, Jmane, Jmui, Jmy37, Jo3sampl, Joanjoc, JobNOB, Jodiejohnson, Joe Schmedley, Joefromrandb, Joey1000, Joeyszy, John Doe or Jane Doe, John Doe42, John M Baker, John254, Jonathansen, Jonik, Jorge Stolfi, JorgeGG, Josefortuna, Jost ammon, Journey118, Jpbowen, Jpers36, Jpo51, Jprg1966, Jrdioko, Jschnur, Jsee, Juliancolton, Jusdafax, Just plain Bill, Justingiovanni, Justlikeus, Jwissick, K1Bond007, KF, Kakofonous, Kamukwam, Karl-Henner, Karol Langner, KatTheFrog, Katanada, Katherine, Kathovo, Katieh5584, Kazushi, Kbenroth, Kbthompson, Kcgiba, Ke4roh, Ke5crz, Kesla, Kevin j, Keyplyr, Kfb2112, Khalid hassani, Khazar2, Kibiusa, Kingkingkingkong, Kingpin13, Kingturtle, Kiore, Kivar2, Kku, Kl4m, Klilidiplomus, Klooka, Kmulholland, KnowledgeOfSelf, Knutux, Koavf, Kociara, Koko333, Kookyunii, Koolkidpaige, Kotra, Koyaanis Qatsi, Krashlandon, Kroussaw, Ku klux clan, Kvdveer, Kwiki, Kyuss-Apollo, L Kensington, Lachatdelarue, Ladwiki, Lal Jensen, Lalascourt21, Lam Kin Keung, Lament, Ldg1364, Leafyplant, LeaveSleaves, Legoktm, LeoDV, Leon..., Leonini, Lerdthenerd, Levineps, Lexistclair, Lexor, Lguzenda, Liamtown, Libraryduffer, Life of Riley, Light Daxter, Light current, Lightmouse, LilHelpa, Limideen, LimoWreck, LinDrug, Little Mountain 5, LittlebutBIG, LiviiBee, Lloydsmith2, Lo2u, Logan, Logical2u, Lordprice, Loren36, LouI, Lradrama, Lucasp1158, Lucasschneid, Luce2809, Luckas Blade, Luckyyy 7, Luna Santin, Lupin, Lyj 88, M.nelson, MASQUERAID, MER-C, MK8, MPerel, Macaddct1984, Macgreco, Macy, Madhero88, Madler, Magioladitis, Magister Mathematicae, MakeRocketGoNow, Mako098765, Mampf-new, Mangostar, Mani1, Manning Bartlett, Manway, MarB4, Marchiond, Marek69, Mark Arsten, Markdeo, Marsman12019, Martin451, Martinkunev, Massacrenz, Master2848, Masterkhufu, Matdrodes, Matewie, Mathboy965, Mathglot, Matt Crypto, Mattgirling, Mattgregg22, Matthew Fennell, Matticus78, Mattjimjapan, Mattley, Mav, Max Naylor, MaxGoldstein, Maxamegalon2000, Maxym98, Mayitoallende, Mayooranathan, Mayumashu, Mazzyroxs, Mbase1235, Mboo101, McSly, Mcmachete, Mdwilliams2, Me123456789ert, Mean as custard, Mee1989, MegX, Meho12345, Mel Etitis, Melsaran, Mentality, Mentifisto, Merapiview, Merphant, Mhking, Miami33139, MichaelRedd22, Michaelas10, Michaelkas9, Microtony, Mikaey, Mike Rosoft, Mikemaehr, Mikeo, Milton Stanley, Mind meal, MinorContributor, Mioanadads, Mipadi, MiroBrada, MissAlyx, Missmarple, Misza13, Mitchnotb1tch, Miwunderlich, Modemac, Modulatum, Moeron, Monkeys rule 99, MonoAV, Monty845, Morahman7vn, MorrisRob, Mpafp, MrTroy, Mrh30, Mrmusichead, Munci, Murphyboy37, Mus Musculus, MusicManSD, Musicindia1, Mwelch, Mwilso24, Mwinog2777, Mwtoews, My3kidz, MyFavoriteMartin, Mygerardromance, Mylyricalsoul, Mütze, N5iln, NBromberg, NHRHS2010, Naddy, Narcisse, NatGeoMusicIntern, Nataev, Natalie Erin, Naval Scene, NawlinWiki, Ndorward, NeatThings, NedKelly95, Neelix, NeilN, Neologisma, Neonblak, Nestabayley, Netkinetic, NewEnglandYankee, NewWaveKid, Newport 63, Niceguyedc, Nicholas Tam, Nick, Nick Drake, Nick UA, NickdelaG, NickelKnowledge, Nickpd4749, NielsenGW, NigelR, Nikai, Nikkimaria, Nikostar, Nishkid64, Nite-Sirk, Nivix, Nocans, Nolatime, Noles99, Nomatey, Nomo, Normandy, Northumbrian, Notjakegyllenhaal, Nsartzetakis, NuclearWarfare, Nuggetmuncher, Nycarmelo, Nyogtha, O.Koslowski, Oda Mari, OfficeGirl, Ohconfucius, Ohnoitsjamie, Ojm, Ollee, Ollie 007, Ollielollie, Olof, Omgwaffels, Omicronpersei8, OnBeyondZebrax, OnePt618, Oneiros, Onelovetema, Opus88888, Orbst, Ortolan88, Ottex, Owen, Oxymoron83, P-we Joplin, PJtP, PL290, PRiis, PS3=awesomeness, PTSE, Pamri, Panamaniac, Paragon12321, Patrick lee-thorp, Paul Armstrong, Paul Richter, Paul-L, PaulHA2, PaulMorel, Paulr, Pax:Vobiscum, Paxsimius, Pbro13, PeaceNT, Pederhorner, Pedro, Pepingen, PerfectStorm, Persian Poet Gal, Peruvianllama, Peter Fleet, Peter Isotalo, Peter Karlsen, Peterdjones, Petersmusic, Petrb, Pfranson, Pgk, PhantomPlugger, PhilKnight, Philip Trueman, Philippe, Piano non troppo, Picapica, Pinball22, Pinethicket, Pishogue, Pj payton, Pjoef, Pjs012915, Plastikspork, Platapus love, Playgalnikki9, Ponyo, Porqin, Poweroid, Prasanta Chowdhury, Pug1776, Pursey, Purslane, Qaddosh, Qero, Quadell, Quercusrobur, Quiddity, Quintote, Qwyrxian, Qxz, Radiofunhouse, Rafamachine, Rag-time4, RaiderTarheel, Rainahlicious, Rainwarrior, Ralphscheider42, Rambosgod, Rancourrancour, RandomAct, Rannpháirtí anaithnid, RattusMaximus, Raul654, Razz1990, Rbmcotterrell, Rd232, Rdelcher, Rdsmith4, Reaper Eternal, Red dwarf, Red-eyed demon, RedWolf, Redheylin,

Redsquirrel118, Redthoreau, Redy to b a man55, Ref, Regibox, Reisio, Rena1998, Res2216firestar, Rettetast, RexNL, Rholton, Rhpt, Rhrad, Rich Farmbrough, RichardMills65, Richger1946, Richiety, Rick Block, Ringbang, Rjstott, Rjwilmsi, Rklawton, Rmhermen, Rmpng, Roadstaa, Robbiewolf, Robert Campagnola, Robert Merkel, RobertG, Robinoke, Robkirwan, Robo Cop, Rock8888, Rocker8222, Rocketrick217, Rockingallnitelong, RocknRollKat, Rocknrollman123, Rocksalt1876, Romanm, Ronhjones, Rory096, Rotisserizer, RoyBoy, Rrburke, Rror, Ruy Pugliesi, Ryan, S02178, SJP, SMC, SRESQ, SWAdair, Saebjorn, Sal Massaro, Saltywood, SamEV, Samsara, Sango123, Sardanaphalus, SarekOfVulcan, Sarferz, Savh, Saxdival, Scarian, Scetoaux, SchfiftyThree, Schiller67, Scoops, ScottSteiner, Scottbell, ScottyBerg, Sct72, Sean mc sean, Seaphoto, Seqsea, Seraphim, SethTisue, Shadowjams, Shaftheking, Shanel, Sharkmandude4000, Shenme, Shoemaker's Holiday, Shoeofdeath, ShotokanTuning, Shsilver, Silence, Silly rabbit, Simmo mania, Singingdaisies, Singularity, SiobhanHansa, Sjakkalle, Sjö, Sketchee, Skidude9950, Skizzik, Skomorokh, SlimVirgin, Sloaner sloan, Slsdig, Sluffs, Sluzzelin, Slysplace, Snidelywiplash, Snigbrook, Snowolf, SoLando, Sobmol, Some jerk on the Internet, Sonez1113, Sophitus, Special-T, Spencer, Spencerb4king, Spicergraham, Spiel496, Spliffy, Squandermania, Squids and Chips, Srikeit, Srnec, Ssamuels, Ssperson, Staeiou, Starber, Starguitar, Station1, Steinizethat, SteloKim, Stephenb, Stevekeiretsu, Steven Zhang, Stevenmitchell, Stirling Newberry, Stuhacking, Styrofoam1994, SuburbaniteFury, Suffusion of Yellow, Summer Song, Sunderland06, Supersportnn, Susurrus, Susvolans, Sutchc, SweetNeo85, Swinglineswingtime, Swpb, Symane, T. Anthony, TBrandley, TK421, TUF-KAT, TachyonJack, Tacolord187, Tar7arus, Tardmaster, Tarquin, Taskerj, Taxman, Tdl1060, Technopat, Teddylevy, Teklund, Template namespace initialisation script, Temporaluser, Tghe-retford, Thatjazzygal6, Thatsloankid, The Anome, The Evil IP address, The Missing Hour, The Rambling Man, The Thing That Should Not Be, The only person ever, TheGrimReaper NS, TheJackBradley, TheKid, TheOldJacobite, Thebenbox, Thecoolguywhos, Thehebrewhammer, Theoldanarchist, Therewillbefact, Thine Antique Pen, Thingg, This lousy T-shirt, Thorltd, Thorpe, Thrissel, Tiddly Tom, Tide rolls, TigerShark, Timrod, Tinton5, Tiptoety, Titoxd, Tleaver, Tlg2000, Tlud, Tolesi, Tolly4bolly, TomPhil, Tomcat7, Tommy The Wise, Tommy2010, Tonsofpcs, Tony1, Toomanylies, Topher385, Tpbradbury, Tpm88, Travelbird, Traxs7, Treisijs, TreveX, Trevor MacInnis, Treyt021, Triwbe, Trontonian, Troy 07, Trusilver, Tsunamishadow, Ttkorhon, Tutmosis, TwilightGirl212, TwoRiversWC, Ufossuck, Uncle Dick, Uriel8, Username1a2b, Utcursch, Vagkoulas1986, Vanished user 39948282, VasilievVV, Velella, Vendettax, Verbist, Verklempt, Versus22, Videmus Omnia, Vina, Visor, Vmhp, Vrenator, Vudujava, Vytal, Wade550, Waggers, Wahabijaz, Walking the blues, Waltoco, Wapo3, WarthogDemon, Wavelength, Wayland, Wayne Miller, Wayne Slam, Wayward, Weedwhacker128, Welling, Wendybaish, Wenli, Whatavividimagination, Whiteboyye, Widr, Wik, WikHead, Wiki alf, Wiki libs, Wikignome0530, Wikigod01, Wikipeditor40, Wikipelli, Wikisarah, Will Beback Auto, Willgomes, Willking1979, Willtron, Wimt, Wine Guy, Wjejskenewr, Wknight94, Wllmlos, Wmahan, Wmpay, Woffie, Wolfwood309, Wombatcat, Woodstock27, Woogee, Woohookitty, Worth my salt, Wrathful mite, Wtmitchell, Wysinger, XXXDemonBoyXXX, Xanchester, Xaosflux, Xaoyuyu, Xhalyt, Xixtas, Xricci, Ydorb, Yid613, Yintan, YouWillBeAssimilated, Youngea, Yromemtnatsisrep, Yuriz, Zabadab, Zabadinho, Zachlipton, Zackomak, Zalgo, Zangorn, Zazaban, Zeppelin4life, Zidonuke, Zoe, Zoicon5, Йованв6, Александър, 3723 anonymous edits

David_Temple *Source*: http://en.wikipedia.org/w/index.php?title=David_Temple *Contributors*: Cjc13, Datemple41, Flapdragon, J04n, JonBradfield, Jpbowen, L888Y5, Mikeconroy39t, Mild Bill Hiccup, Plasticup, Sallyjaynehall, Wordmusicimage, 4 anonymous edits

Buxton_Festival *Source*: http://en.wikipedia.org/w/index.php?title=Buxton_Festival *Contributors*: Anonymous101, Bearcat, BrainyBabe, Colonies Chris, Dave.Dunford, Debbieporter, Grstain, J.P.Lon, JForget, JIP, Johnbod, Mattyk, Mowsbury, Phoe, Ssilvers, Tom crawshaw, Trident13, Violncello, Viva-Verdi, WJBscribe, 24 anonymous edits

Image Sources, Licenses and Contributors

File:Classical music composers montage.JPG *Source*: http://en.wikipedia.org/w/index.php?title=File:Classical_music_composers_montage.JPG *License*: unknown *Contributors*: User:Clestur, User:Nihil novi, User:Rodin777, User:אמסיטה123

File:Tch In Charlotte.JPG *Source*: http://en.wikipedia.org/w/index.php?title=File:Tch_In_Charlotte.JPG *License*: unknown *Contributors*: User:Derek Gleeson

File:Vielle.jpg *Source*: http://en.wikipedia.org/w/index.php?title=File:Vielle.jpg *License*: unknown *Contributors*: Aotake, Fransvannes, Mattes, Wilimut, Wst, 1 anonymous edits

File:CordierColor.jpg *Source*: http://en.wikipedia.org/w/index.php?title=File:CordierColor.jpg *License*: unknown *Contributors*: Baude Cordier

File:white mensural notation.gif *Source*: http://en.wikipedia.org/w/index.php?title=File:White_mensural_notation.gif *License*: unknown *Contributors*: Aviad2001, Cdang, Maksim, Wst

File:Baschenis - Musical Instruments.jpg *Source*: http://en.wikipedia.org/w/index.php?title=File:Baschenis_-_Musical_Instruments.jpg *License*: unknown *Contributors*: Aotake, Bukk, Clusternote, Darwinius, G.dallorto, Harpsichord246, Hsarrazin, Mattes, Phrood, RIbberlin, Skipjack, Valérie75, Warburg

File:Haydnportrait.jpg *Source*: http://en.wikipedia.org/w/index.php?title=File:Haydnportrait.jpg *License*: unknown *Contributors*: Photo Nevilley at en.wikipedia

File:Stravinsky picasso.png *Source*: http://en.wikipedia.org/w/index.php?title=File:Stravinsky_picasso.png *License*: unknown *Contributors*: Awops, Henry Flower, Lupin, Mathematiks, Mechamind90, Orionist, Quadell, Robert.Allen, Shyam, Thuresson, 1 anonymous edits

File:Louis-Nicolas Clerambault.jpg *Source*: http://en.wikipedia.org/w/index.php?title=File:Louis-Nicolas_Clerambault.jpg *License*: unknown *Contributors*: Aotake, Calliopejen1, Ecummenic, Frinck51, Mu, Shakko

File:Temporal-Control-and-Hand-Movement-Efficiency-in-Skilled-Music-Performance-pone.0050901.s001.ogv *Source*: http://en.wikipedia.org/w/index.php?title=File:Temporal-Control-and-Hand-Movement-Efficiency-in-Skilled-Music-Performance-pone.0050901.s001.ogv *License*: unknown *Contributors*: Goebl W, Palmer C

File:Palace Green Library.jpg *Source*: http://en.wikipedia.org/w/index.php?title=File:Palace_Green_Library.jpg *License*: unknown *Contributors*: Gaspa

File:Durham Cathedral from Durham School.jpg *Source*: http://en.wikipedia.org/w/index.php?title=File:Durham_Cathedral_from_Durham_School.jpg *License*: unknown *Contributors*: User:Teach46

File:Durham School from the road.jpg *Source*: http://en.wikipedia.org/w/index.php?title=File:Durham_School_from_the_road.jpg *License*: unknown *Contributors*: User:Teach46

File:Durham School - chapel.jpg *Source*: http://en.wikipedia.org/w/index.php?title=File:Durham_School_-_chapel.jpg *License*: unknown *Contributors*: User:Teach46

File:Durham School - chapel steps.jpg *Source*: http://en.wikipedia.org/w/index.php?title=File:Durham_School_-_chapel_steps.jpg *License*: unknown *Contributors*: User:Teach46

File:Durham School - chapel closeup.jpg *Source*: http://en.wikipedia.org/w/index.php?title=File:Durham_School_-_chapel_closeup.jpg *License*: unknown *Contributors*: User:Teach46

File:Durham School - all weather pitch.jpg *Source*: http://en.wikipedia.org/w/index.php?title=File:Durham_School_-_all_weather_pitch.jpg *License*: unknown *Contributors*: User:Teach46

File:university of bristol tower after cleaning arp.jpg *Source*: http://en.wikipedia.org/w/index.php?title=File:University_of_bristol_tower_after_cleaning_arp.jpg *License*: unknown *Contributors*: Arpingstone

File:King's College London Students Evacuated To Bristol, England, 1940 D433.jpg *Source*: http://en.wikipedia.org/w/index.php?title=File:King's_College_London_Students_Evacuated_To_Bristol,_England,_1940_D433.jpg *License*: unknown *Contributors*: Fæ, Man vyi

File:Victoria Rooms (750px).jpg *Source*: http://en.wikipedia.org/w/index.php?title=File:Victoria_Rooms_(750px).jpg *License*: unknown *Contributors*: Original uploader was Rbrwr at en.wikipedia

File:Bristol University from Cabot Tower.jpg *Source*: http://en.wikipedia.org/w/index.php?title=File:Bristol_University_from_Cabot_Tower.jpg *License*: unknown *Contributors*: User:Arpingstone

File:QEH Prize Day 2004.JPG *Source*: http://en.wikipedia.org/w/index.php?title=File:QEH_Prize_Day_2004.JPG *License*: unknown *Contributors*: Mike Buck

File:Goldney2.jpg *Source*: http://en.wikipedia.org/w/index.php?title=File:Goldney2.jpg *License*: unknown *Contributors*: User:Francium12

File:Geography Department, University of Bristol - geograph.org.uk - 201159.jpg *Source*: http://en.wikipedia.org/w/index.php?title=File:Geography_Department,_University_of_Bristol_-_geograph.org.uk_-_201159.jpg *License*: unknown *Contributors*: The wub

File:University of Bristol buildings.JPG *Source*: http://en.wikipedia.org/w/index.php?title=File:University_of_Bristol_buildings.JPG *License*: unknown *Contributors*: Francium12 (talk) Original uploader was Francium12 at en.wikipedia

File:Wills memorial library.JPG *Source*: http://en.wikipedia.org/w/index.php?title=File:Wills_memorial_library.JPG *License*: unknown *Contributors*: Milla and Mark

File:UniversityofBristolEngineering.JPG *Source*: http://en.wikipedia.org/w/index.php?title=File:UniversityofBristolEngineering.JPG *License*: unknown *Contributors*: Francium12 (talk) Original uploader was Francium12 at en.wikipedia

File:University of Bristol, School of Chemistry.jpg *Source*: http://en.wikipedia.org/w/index.php?title=File:University_of_Bristol,_School_of_Chemistry.jpg *License*: unknown *Contributors*: User:Rwendland

File:070327 ukbris ubu01.jpg *Source*: http://en.wikipedia.org/w/index.php?title=File:070327_ukbris_ubu01.jpg *License*: unknown *Contributors*: User:Steinsky

File:Wills Hall, University of Bristol.JPG *Source*: http://en.wikipedia.org/w/index.php?title=File:Wills_Hall,_University_of_Bristol.JPG *License*: unknown *Contributors*: Tim giddings at en.wikipedia

File:Goldney1.jpg *Source*: http://en.wikipedia.org/w/index.php?title=File:Goldney1.jpg *License*: unknown *Contributors*: User:Francium12

File:CliftonHillHouse.JPG *Source*: http://en.wikipedia.org/w/index.php?title=File:CliftonHillHouse.JPG *License*: unknown *Contributors*: Francium12 (talk) Original uploader was Francium12 at en.wikipedia

File:UniversityofBristollogo2.jpg *Source*: http://en.wikipedia.org/w/index.php?title=File:UniversityofBristollogo2.jpg *License*: unknown *Contributors*: Francium12 (talk) Original uploader was Francium12 at en.wikipedia

File:PharoahSanders.jpg *Source*: http://en.wikipedia.org/w/index.php?title=File:PharoahSanders.jpg *License*: unknown *Contributors*: Dr. Shaggeman, Infrogmation, Mats Halldin, Mind meal, Ras67, Wikielwikingo, Wikiwal, 3 anonymous edits

Image:Congo-early.gif *Source*: http://en.wikipedia.org/w/index.php?title=File:Congo-early.gif *License*: unknown *Contributors*: E. W. Kemble

File:Slave dance to banjo, 1780s.jpg *Source*: http://en.wikipedia.org/w/index.php?title=File:Slave_dance_to_banjo,_1780s.jpg *License*: unknown *Contributors*: AnRo0002, Calliopejen, Cirt, Dave souza, Guitarpop, Infrogmation, Quibik, 2 anonymous edits

File:Virginia Minstrels, 1843.jpg *Source*: http://en.wikipedia.org/w/index.php?title=File:Virginia_Minstrels,_1843.jpg *License*: unknown *Contributors*: Alton, Amcaja, Badseed, G.dallorto, Infrogmation, 2 anonymous edits

File:Tresillo divisive.png *Source*: http://en.wikipedia.org/w/index.php?title=File:Tresillo_divisive.png *License*: unknown *Contributors*: User:Dr clave

File:Candombe1870-Uruguay.jpg *Source*: http://en.wikipedia.org/w/index.php?title=File:Candombe1870-Uruguay.jpg *License*: unknown *Contributors*: Fadesga, Mattes, Negromacondo, Photosbal

File:Tresillo+ backbeat.jpg *Source*: http://en.wikipedia.org/w/index.php?title=File:Tresillo+_backbeat.jpg *License*: unknown *Contributors*: User:Dr clave

File:Cinquello bell pattern.png *Source*: http://en.wikipedia.org/w/index.php?title=File:Cinquello_bell_pattern.png *License*: unknown *Contributors*: User:Dr clave

File:Scott Joplin 19072.jpg *Source*: http://en.wikipedia.org/w/index.php?title=File:Scott_Joplin_19072.jpg *License*: unknown *Contributors*: Durova, Illegitimate Barrister, Major Bloodnok

Image:Maple Leaf Rag seventh chord resolution.png *Source*: http://en.wikipedia.org/w/index.php?title=File:Maple_Leaf_Rag_seventh_chord_resolution.png *License*: unknown *Contributors*: User:SreeBot

File:Solace.tiff *Source*: http://en.wikipedia.org/w/index.php?title=File:Solace.tiff *License*: unknown *Contributors*: User:Dr clave

File:Blues and pentatonic scales..jpg *Source*: http://en.wikipedia.org/w/index.php?title=File:Blues_and_pentatonic_scales..jpg *License*: unknown *Contributors*: User:Dr clave

File:Bolden band.gif *Source*: http://en.wikipedia.org/w/index.php?title=File:Bolden_band.gif *License*: unknown *Contributors*: Cirt, Hpschaefer, Infrogmation, Thuresson, 1 anonymous edits

File:Big four Buddy Bolden.png *Source*: http://en.wikipedia.org/w/index.php?title=File:Big_four_Buddy_Bolden.png *License*: unknown *Contributors*: Dr clave (talk)

File:Jelly Roll Blues 1915.jpg *Source*: http://en.wikipedia.org/w/index.php?title=File:Jelly_Roll_Blues_1915.jpg *License*: unknown *Contributors*: Publisher: Will Rossiter

File:New orleans blues corrected.jpg *Source*: http://en.wikipedia.org/w/index.php?title=File:New_orleans_blues_corrected.jpg *License*: unknown *Contributors*: User:Dr clave

File:Swing pattern in relation to straight subdivisions.jpg *Source*: http://en.wikipedia.org/w/index.php?title=File:Swing_pattern_in_relation_to_straight_subdivisions.jpg *License*: unknown *Contributors*: User:Dr clave

File:WC Handy age 19 handyphoto10.jpg *Source*: http://en.wikipedia.org/w/index.php?title=File:WC_Handy_age_19_handyphoto10.jpg *License*: unknown *Contributors*: uploaded to Wikipedia by Mink Butler Davenport

File:St louis blues.tif *Source*: http://en.wikipedia.org/w/index.php?title=File:St_louis_blues.tif *License*: unknown *Contributors*: User:Dr clave

File:Louis Armstrong restored.jpg *Source*: http://en.wikipedia.org/w/index.php?title=File:Louis_Armstrong_restored.jpg *License*: unknown *Contributors*: World-Telegram staff photographer

File:Jazzing orchestra 1921.png *Source*: http://en.wikipedia.org/w/index.php?title=File:Jazzing_orchestra_1921.png *License*: unknown *Contributors*: Clusternote, Dr. Shaggeman, Infrogmation, Lobo, Magister Mathematicae, Solkoll, TDogg310, 1 anonymous edits

File:Mandy (louis armstrong).tiff *Source*: http://en.wikipedia.org/w/index.php?title=File:Mandy_(louis_armstrong).tiff *License*: unknown *Contributors*: User:Dr clave

File:BennyGoodmanStageDoorCanteen.jpg *Source*: http://en.wikipedia.org/w/index.php?title=File:BennyGoodmanStageDoorCanteen.jpg *License*: unknown *Contributors*: Film screenshot

File:Duke Ellington at the Hurricane Club 1943.jpg *Source*: http://en.wikipedia.org/w/index.php?title=File:Duke_Ellington_at_the_Hurricane_Club_1943.jpg *License*: unknown *Contributors*: AnRo0002, Dodo, Infrogmation, Mike Dillon

File:Thelonious Monk, Minton's Playhouse, New York, N.Y., ca. Sept. 1947 (William P. Gottlieb 06191).jpg *Source*: http://en.wikipedia.org/w/index.php?title=File:Thelonious_Monk,_Minton's_Playhouse,_New_York,_N.Y.,_ca._Sept._1947_(William_P._Gottlieb_06191).jpg *License*: unknown *Contributors*: Freimut Bahlo, Infrogmation, Ras67

File:Earl Hines 1947.jpg *Source*: http://en.wikipedia.org/w/index.php?title=File:Earl_Hines_1947.jpg *License*: unknown *Contributors*: Cherubino, Freimut Bahlo

File:Charlie Parker, Tommy Potter, Miles Davis, Max Roach (Gottlieb 06941).jpg *Source*: http://en.wikipedia.org/w/index.php?title=File:Charlie_Parker,_Tommy_Potter,_Miles_Davis,_Max_Roach_(Gottlieb_06941).jpg *License*: unknown *Contributors*: Cherubino, Ras67, 1 anonymous edits

File:Machito and his sister Graciella Grillo.jpg *Source*: http://en.wikipedia.org/w/index.php?title=File:Machito_and_his_sister_Graciella_Grillo.jpg *License*: unknown *Contributors*: Clusternote, Ethanpolonsky, Lioneldecoster

File:Kpanlogo bell.jpg *Source*: http://en.wikipedia.org/w/index.php?title=File:Kpanlogo_bell.jpg *License*: unknown *Contributors*: User:Dr clave

File:Dizzy Gillespie playing horn 1955.jpg *Source*: http://en.wikipedia.org/w/index.php?title=File:Dizzy_Gillespie_playing_horn_1955.jpg *License*: unknown *Contributors*: Ary29, Dcoetzee, Docu, Frank C. Müller, G.dallorto, Infrogmation, Jkelly, Jossifresco, Juiced lemon, Mjrmtg, Romary, Wst

File:Manteca = bell pattern.tif *Source*: http://en.wikipedia.org/w/index.php?title=File:Manteca_=_bell_pattern.tif *License*: unknown *Contributors*: User:Dr clave

File:Mongo Santamaria 1969.JPG *Source*: http://en.wikipedia.org/w/index.php?title=File:Mongo_Santamaria_1969.JPG *License*: unknown *Contributors*: Associated Booking Corporation (management); photographer: Ronnie Brathwaite

File:Afro blue bass.jpg *Source*: http://en.wikipedia.org/w/index.php?title=File:Afro_blue_bass.jpg *License*: unknown *Contributors*: User:Dr clave

File:So what changes.tif *Source*: http://en.wikipedia.org/w/index.php?title=File:So_what_changes.tif *License*: unknown *Contributors*: User:Dr clave

File:Peter-broetzmann.jpg *Source*: http://en.wikipedia.org/w/index.php?title=File:Peter-broetzmann.jpg *License*: unknown *Contributors*: user:nomo

File:Naná Vasconcelos.jpg *Source*: http://en.wikipedia.org/w/index.php?title=File:Naná_Vasconcelos.jpg *License*: unknown *Contributors*: Gerardo Lazzari

File:Randy Weston.jpg *Source*: http://en.wikipedia.org/w/index.php?title=File:Randy_Weston.jpg *License*: unknown *Contributors*: Bob Travis Casco, Maine, USA

File:Willie bobo afro blue.tif *Source*: http://en.wikipedia.org/w/index.php?title=File:Willie_bobo_afro_blue.tif *License*: unknown *Contributors*: User:Dr clave

File:FOOTPRINTS BASS LINES.tif *Source*: http://en.wikipedia.org/w/index.php?title=File:FOOTPRINTS_BASS_LINES.tif *License*: unknown *Contributors*: User:Dr clave

File:"African Queen" Joe Henderson excerpt.tiff *Source*: http://en.wikipedia.org/w/index.php?title=File:"African_Queen"_Joe_Henderson_excerpt.tiff *License*: unknown *Contributors*: User:Dr clave

File:I IV V pentatonic.tiff *Source*: http://en.wikipedia.org/w/index.php?title=File:I_IV_V_pentatonic.tiff *License*: unknown *Contributors*: User:Dr clave

File:II V I.tiff *Source*: http://en.wikipedia.org/w/index.php?title=File:II_V_I.tiff *License*: unknown *Contributors*: User:Dr clave

File:Miles Davis 24.jpg *Source*: http://en.wikipedia.org/w/index.php?title=File:Miles_Davis_24.jpg *License*: unknown *Contributors*: Oliver Nurock @ ohjaygee Johannesburg/ Cape Town, South Africa

File:Wynton Marsalis 2009 09 13.jpg *Source*: http://en.wikipedia.org/w/index.php?title=File:Wynton_Marsalis_2009_09_13.jpg *License*: unknown *Contributors*: Eric Delmar. Original uploader was Bwalko at en.wikipedia

File:David Sanborn 2008 2.jpg *Source*: http://en.wikipedia.org/w/index.php?title=File:David_Sanborn_2008_2.jpg *License*: unknown *Contributors*: Noticaribe

File:Gangstarr-02-mika.jpg *Source*: http://en.wikipedia.org/w/index.php?title=File:Gangstarr-02-mika.jpg *License*: unknown *Contributors*: User:Mika-photography

File:John Zorn.jpg *Source*: http://en.wikipedia.org/w/index.php?title=File:John_Zorn.jpg *License*: unknown *Contributors*: Andy Newcombe Farnborough, UK

Image:Steve Coleman 1611.JPG *Source*: http://en.wikipedia.org/w/index.php?title=File:Steve_Coleman_1611.JPG *License*: unknown *Contributors*: Mbase1235, Mind meal, NobbiP, Shoulder-synth, Tica, Túrelio, 2 anonymous edits

Image:Buxton Opera House Sign.jpg *Source*: http://en.wikipedia.org/w/index.php?title=File:Buxton_Opera_House_Sign.jpg *License*: unknown *Contributors*: Ian from Buxton, United Kingdom

MIX
Papier aus verantwortungsvollen Quellen
Paper from responsible sources
FSC® C105338

Printed by Books on Demand GmbH, Norderstedt / Germany